BOEING 747

1970 onwards (all marks)

ZENITH PRESS

Haynes

Copyright © 2012 Chris Wood

Published in North America in 2012 by Zenith Press,
an imprint of Quayside Publishing Group,
400 1st Avenue North, Suite 300, Minneapolis,
MN 55401, USA, by arrangement with
Haynes Publishing.

Chris Wood has asserted his moral right to
be identified as the author of this work.

Zenith Press titles are also available at discounts in
bulk quantity for industrial or sales-promotional use.
For details write to Special Sales Manager at
Quayside Publishing Group, 400 1st Avenue North,
Suite 300, Minneapolis, MN 55401, USA.

To find out more about our books, join us online at
www.zenithpress.com or www.qbookshop.com.

ISBN-13: 978-0-7603-4293-0

While every effort is taken to ensure the accuracy
of the information given in this book, no liability
can be accepted by the author or publishers for
any loss, damage or injury caused by errors in, or
omissions from, the information given.

Printed in the USA by Odcombe Press LP,
1299 Bridgestone Parkway, LaVergne, TN 37086.

BOEING 747

1970 onwards (all marks)

Zenith Press

Haynes

Owners' Workshop Manual

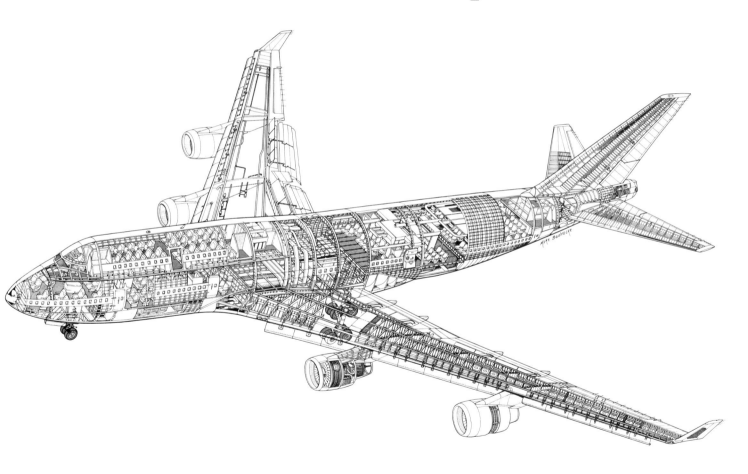

An insight into owning, flying, and
maintaining the iconic jumbo jet

Chris Wood

Contents

OPPOSITE **Assembly work on the nose section of a 747-400 inside Boeing's Everett plant at Seattle. The Everett assembly facility is the world's largest building by volume.**
(Larry MacDougal/PA Images)

Introduction

Since the birth of manned, powered flight a little over 100 years ago, there have been a number of iconic aircraft that have defined a generation or an era. As far as commercial passenger aircraft go, the Douglas DC-3 and the Anglo/French Concorde spring to mind, but the Boeing 747 has to be up there with the greats, both in terms of its size and its effect on the world in which we live.

OPPOSITE A familiar sight at airports around the world for more than 40 years, the arrival of the Boeing 747 on the air transport scene in 1970 was one of the great defining moments in aviation history. This is British Airways 747-400, G-CIVT, preparing to depart from Las Vegas on 4 March 2012.

The first of the wide-body jets, the Boeing 747 brought air travel to the masses on a truly global scale. Boeing estimated that within six months of entering service in January 1970 the existing 747 fleet, by that stage already operated by 30 airlines, had carried its millionth passenger. They have now built over 1,400 747s, they are still building them, and the aircraft is now in its fifth decade of service with the world's airlines. The number of passengers carried by 747s must now run into billions and, whilst the -400 version is approaching its twilight years, the third generation of the 747 is starting to enter service.

This book takes a look at the second generation aircraft, the 747-400, which happens to be the best-selling version (so far). It deals predominantly with the standard passenger version, powered by General Electric engines, which also transpires to be the best-selling engine on the -400 model. Whilst the 747-400 is in many ways not that different to its earlier sibling (and is still very much a 747 at heart), in many ways it is also completely different. The aim of this book is to give the reader a look

under its skin and provide an insight into what makes it tick, plus a few tips on how to fly it as well as how to maintain it.

It is a complex machine – Boeing claims it has six million parts – and the space available in these pages allows for far fewer words than that. You could easily fill several bookcases with information about the 747! This book offers, therefore, an insight into the aircraft and its systems, and it cannot hope to cover every detail. There are many devices mentioned in the text but limited space does not permit an explanation of what they are or how they work. However, we do live in the internet age so if you do not know what a rotary variable differential transformer is, then a well-known internet search engine can probably help!

Not all aircraft of the same model are actually identical; there are different options available when they are purchased and there are often modifications made to them during their working lives which, for an aircraft like the 747, can be 25 to 30 years. Each operator has different ways of doing the same thing and the procedures for operating them also evolve over time, so if the reader is familiar with the 747-400 they may well notice discrepancies with the aircraft they know.

It should also be stressed that this book is for general information only and if the reader is involved in operating 747s, it is *not* a substitute for the Boeing or operator-provided and approved manuals.

The author is an experienced 747 pilot, who has spent the last 20 years flying both the first generation 'Classic' 747 and the second generation -400, and he will soon be entering his third decade of flying this amazing and awe-inspiring leviathan of the air. Even so, he has learnt a lot whilst preparing this book!

RIGHT Global Supply Systems Boeing 747-8F, G-GSSE, at Houston, Texas, on 4 March 2012. *(Mark Parsons)*

BELOW The landing gear is almost retracted as this Virgin 747-400, G-VROY, climbs away from Las Vegas, March 2012.

Chapter One

The Boeing 747 Story

———●———

When the first Boeing 747 touched down at Heathrow Airport from New York on 22 January 1970, it heralded the beginning of a popular revolution in long-haul air travel. Bigger than any passenger aircraft that had come before it, the wide-body 747 'jumbo jet' enabled airlines to slash ticket costs, which meant that affordable travel to faraway destinations finally came within reach of the masses.

OPPOSITE Exemplifying the versatility of the basic Boeing 747 design, this specialist freighter conversion from a passenger -400 series is operated by Air China Cargo. The 747-400BDSF is a conversion by Israel Aerospace Industries, the BDSF standing for 'Bedek Special Freighter'. It can carry 80 tons of cargo. This is 747-400BDSF, B-2478, departing Ted Stevens Anchorage International Airport, Alaska, on 18 April 2008. Air China Cargo is jointly owned by Air China and Cathay Pacific. In 2012 it was operating ten Boeing 747-400 cargo aircraft. *(Karl Drage)*

Conceived, designed and built during the 1960s, the Boeing 747 was born into a decade of rapid growth in air travel and major technological advances in aviation. Credit for creating the 747 goes primarily to three men:

Juan Trippe the Chairman of Pan American World Airways, who saw the market for a new large aircraft; Bill Allen, President of the Boeing Airplane Company, who committed to build an aircraft to fulfil Trippe's vision; and Boeing engineer Joe Sutter who led the design team that created it.

The 747 was the fourth aircraft in Boeing's family of jet airliners, following on from their 707, 727 and 737 models (the 717 was a number originally given to a military derivative of the 707 design for the United States Air Force (USAF), more commonly known as the C-135. Following the merger with McDonnell Douglas in 1997 it was subsequently applied to that company's MD-95).

The 747's roots can be traced back to Boeing's first jet aircraft, the B-47 Stratojet bomber, which was designed for the United States Army Air Force (USAAF) as a replacement for its Second World War era piston-engine bombers, such as Boeing's B-29 Superfortress. The B-47 was a radical new design that benefitted from two key facts.

First was Boeing's decision in August 1941 to build its own wind tunnel, and not just to build a tunnel but to make it a high-speed one even

ABOVE Stablemates of the Boeing 747 in the manufacturer's 7-series of jet airliners were the Boeing 727 (illustrated is 727-200, N413DA, at Boston on 30 August 1998) and ...

LEFT ... Boeing 737 (737-200, ZS-OLB, pictured at Johannesburg on 2 December 2010). *(Chris Wood)*

LEFT The B-47 Stratojet's performance was breathtaking and far surpassed that of its straight-wing contemporaries. Its podded engines and swept wings were significant influences on modern jet airliner design. This B-47 is taking off with RATO (Rocket Assisted Take-Off). *(Boeing)*

though there were no high-speed aeroplanes on the drawing board. Having their own tunnel allowed Boeing unrestricted use of this resource and it also meant that all the data could be kept confidential. Prior to building their own, Boeing had to share the use of wind tunnels with other aircraft manufacturers, such as the one at the California Institute of Technology (Caltech) in Pasadena near Los Angeles.

The second was the discovery of German research data into high-speed flight. As the Second World War was drawing to a close in Europe the Americans sent a team of scientists and engineers to Germany to investigate the Nazi regime's technological advances in aeronautics. This team was led by the noted aerodynamicist Theodore von Karmen, the Director of the Guggenheim Aeronautical Laboratory at Caltech; Boeing's chief aerodynamicist George Schairer was a member of the team. At a secret Aeronautics Research Laboratory at Volkenrode, east of Hannover, they discovered wind tunnel data that showed the benefits of swept wings for flight at speeds close to the speed of sound. Schairer sent details of this research back to Boeing.

With the war almost over, the USAAF was planning for the future and looking at uses for the new jet engines that were becoming available. One project was for a new multi-engine jet bomber, and four manufacturers were contracted to begin design studies – North American with their XB-45, Convair with the XB-46, Boeing with their XB-47 and Martin with their XB-48. Northrop was also developing a bomber with their radical YB-49 flying wing. Most early designs of jet aircraft had followed on from their piston-engine predecessors by having straight wings and engines mounted either in the fuselage or embedded in the wings. America's first jet, the Bell P-59 Airacomet had its engines mounted in the fuselage, and also featured straight wings. The world's second operational jet fighter, Britain's Gloster Meteor, had its two engines mounted in the middle of its straight wings.

The first US jet bomber to fly had actually been the Douglas XB-43, which had a straight, high wing and two engines mounted in the fuselage. The XB-45, 46 and 48 all followed this doctrine, with straight wings and engines mounted in or just under the wings. Boeing's design had started off in that way and it went through several incarnations before the final layout was decided upon. Of the other designs, only the B-45 entered production and only 143 were built.

BELOW Britain's Gloster Meteor was the Allies' first (but the world's second) operational jet fighter, which had its twin turbojets mounted mid-span in each wing. This is an NF11, the prototype of which first flew in 1950. *(Chris Wood)*

Interestingly, and perhaps not surprisingly, some early German jet aircraft designs such as the Messerschmitt Me262, which was the world's first operational jet fighter, had a modestly swept wing, and also featured its engines slung under the wing.

The USAAF was not happy with having jet engines embedded in the fuselage or the wings of large aircraft as they were concerned about the effects that engine fires and battle damage could have on its survivability. Boeing's designers came up with the idea of mounting the engines of the B-47 in pods slung under the wings. This had several major advantages:

- The weight of the engines mounted on the outboard parts of the wing helped to dampen any flexing.
- It made the wing aerodynamically cleaner so therefore more efficient.
- It made the engines easy to access for maintenance purposes.
- It made it relatively easy to change the type of engine.
- It moved the engine noise away from the fuselage.
- It meant that any engine fires or catastrophic failures were unlikely to compromise the structural integrity of the wing.

As a direct result of the discovery at Volkenrode, and following extensive testing in their own wind tunnel, Boeing redesigned their new jet-engine bomber with a wing swept back at 35°.

To satisfy the USAAF's requirement the engines were in pods slung beneath the wings.

Having the engines in under-slung pods also allowed the B-47 to have a very thin wing, which was good aerodynamically. However, it left nowhere to mount the landing gear. The solution Boeing came up with was to fit bicycle main landing gear in the fuselage, with outrigger wheels near the wingtips. To manoeuvre the aircraft on the ground the front wheels were steerable. This

ABOVE Aerospace design technology captured in Germany at the end of the Second World War meant the US aircraft industry could benefit from the Germans' extensive experiments into swept-wing designs, like the Messerschmitt Me262 jet fighter-bomber seen here. *(Jackie Nudd)*

LEFT Using Nazi swept-wing design and a tricycle undercarriage configuration, the Boeing B-47 Stratojet was Strategic Air Command's (SAC) first swept-wing jet bomber when it entered USAF service in 1951. *(Boeing)*

system was powered hydraulically and controlled through the rudder pedals.

The prototype B-47 made its first flight on 17 December 1947, the 44th anniversary of the Wright brother's first manned powered flight. However, flight-testing of the B-47 showed up some undesirable handling characteristics. The aircraft had a tendency to Dutch Roll, a phenomenon known about since the early days of flight but which was much more prevalent on a swept-wing aircraft. No solution was then available but in the space of three weeks Boeing's engineers came up with a device known as the yaw damper, which was fitted to the prototype B-47 and solved the problem.

The B-47's wings were designed to be flexible to cope with the bending stresses experienced during flight. However, during a high-speed test an attempt to turn right resulted in the aircraft turning left. Investigation showed that above a particular speed, applications of aileron resulted in the wing starting to twist, and the effect of this twist was overriding the aileron input resulting in control reversal. Boeing's solution to this problem was to fit spoiler panels in the upper surface of the wing (these had been a feature on Convair's XB-46). When deployed these panels 'spoil' the lift on their respective wing so aiding turning in the desired direction.

Swept wings, yaw dampers and spoilers are to be found on all large jets, and the vast majority of jet airliners also have their engines mounted in pods under the wing. (Production B-47s didn't have spoilers as they were considered unnecessary, the solution was to restrict the aircraft's maximum speed instead.)

Hot on the heels of the B-47 came Boeing's next offering, its giant B-52 Stratofortress, which was first flown on 15 April 1952 by the legendary test pilot Alvin 'Tex' Johnston (it is coincidence that the B-47 first flew in 19**47**, and the B-52 in 19**52**!). Having started out as a turboprop with a modest amount of wing sweep, the B-52 was redesigned with a wing swept to 35°, podded jet engines and the main landing gear mounted in the fuselage. Innovations in this design included the use of hot air bled from the engines to power some services such as the hydraulic pumps. (Unlike the B-47, the B-52 did have spoilers, and the later versions had no ailerons!) The other innovation was the use of subcontracting, with ultimately over half of the aircraft structure being built by companies other than Boeing.

Whilst busy building jet bombers for the USAF (which had split away from the Army to become a separate air force in September 1947), Boeing was also turning its attention to the possibilities of jet-powered transport aircraft. The pre-eminent builder of airliners in the United States at the time was Douglas in California, which had built the ubiquitous DC-3 and followed it up with the DC-4 and DC-6 piston-engine aircraft. These had sold in large numbers and Douglas was busy developing a follow-up to the DC-6 in the shape of the DC-7, so it was not particularly interested in jets. Also in California was Lockheed, whose offering for the airliner market was its elegant Constellation, also piston-powered. Boeing, meanwhile, had developed a version of its B-29 Superfortress, the Boeing model 367 Stratofreighter, as a military freighter (designated C-97 by the USAF) and had subsequently produced a civil version, the model 377 Stratocruiser. This was a luxurious long-range airliner but was plagued with problems, particularly with its engines, and only sold in limited numbers. It first flew on 8 July 1947, just five months before the B-47. A young engineer by the name of Joe Sutter,

RIGHT 'Tex' Johnston became a test pilot for Boeing in 1948. He achieved a certain notoriety on 7 August 1955 when he barrel-rolled Boeing's 367-80 experimental jet (the Dash-80, forerunner of the 707) on a demonstration flight near Seattle. *(Boeing)*

recently hired by Boeing, had been put to work on the Stratocruiser.

Boeing saw the need for a jet powered tanker for the USAF to refuel its new fleet of B-52s. At the time the USAF's tanker was the KC-97 (a version of the piston-engine C-97), which was fine for refuelling the USAF's B-50s (a modified version of the B-29) but was not ideal for refuelling jets, as it could not fly anywhere near as high or as fast.

On 22 April 1952, just a week after the first flight of the B-52, Boeing made the bold decision to invest its own money (it eventually cost US $16 million, although it is claimed that some of the cost was funded by the Government under 'Independent Research and Development') to develop a prototype jet transport. Hedging its bets, Boeing designed it with the potential to be used as a military tanker,

ABOVE Production of the B-52 ended in 1962 after 744 examples were built, culminating in the B-52G and H versions. B-52s saw action in the bomber and reconnaissance roles in Vietnam, both Gulf Wars and over Afghanistan. In 2012 around 80 B-52H models remain in service with the USAF. This is B-52H, 61-0013, climbing out of Nellis AFB, Nevada, on 25 August 2009. *(Chris Wood)*

LEFT Boeing's B367-80 prototype jet transport, conceived as a 'proof of concept vehicle', was powered by the same engine as the B-52, the Pratt & Whitney J-57 turbojet. It also shared common design characteristics with the big bomber (in the background of this photograph) like the podded engines and swept wings. The B367-80 became known simply as the Dash 80 when it was unveiled on 15 May 1954. *(PRM Aviation)*

a military freighter or a commercial airliner. The aircraft was to be a proof of concept vehicle, rather than an actual prototype, and was given the designation 367-80. It was to be powered by the same engine as the B-52, the Pratt & Whitney J-57 turbojet. The design went through many arrangements based on and around the Stratocruiser, the B-47 and the B-52 before the final configuration was drawn up. Perhaps not surprisingly this aircraft featured a wing swept at 35° and its jet engines in pods slung under the wings. It had tricycle landing gear, with twin nose wheels and two four-wheel main gear bogies, which retracted sideways into the wing root and fuselage.

The world's first jet airliner, Britain's de Havilland Comet, had wings with a modest amount of sweep, which restricted its cruising speed, and engines embedded in the wing root.

The 367-80 featured a host of other innovations. For roll control it was fitted with two sets of ailerons, a high-speed set and a low-speed set. The low-speed ailerons were only available when the flaps were lowered; when the flaps were retracted they were locked into position. The spoilers had been modified so that they not only aided roll control, they could also be used as airbrakes in the air and on the ground. Deploying the spoilers on landing dumps the lift from the wing, ensuring the aircraft weight is on the wheels, so aiding braking. To assist stopping on the ground it had clamshell thrust reversers on its engines.

The Dash 80, as it became known, was rolled out on 15 May 1954 and first flew two months later on 15 July, piloted by 'Tex' Johnston.

From this proof of concept vehicle Boeing went on to develop the KC-135 tanker for the USAF and the 707 for airline use. In 1954 the USAF ordered its first batch of KC-135s and on 13 October 1955 Pan American World Airways placed the first order for the 707, ordering 20. However, by this time Douglas were catching up and had its own four-engine jet under development, the DC-8, and Pan Am also ordered 25 of these.

The first 707 flew on 20 December 1957, again in the hands of 'Tex' Johnson. Pan Am was the first to put it into service, the first flight being from New York to Paris on 26 October 1958.

Initial aircraft were powered by Pratt &

Whitney JT-3C turbojets (a version of the J-57 military turbojet) and the more powerful JT-4A (a version of the J-75). Subsequently a turbofan version of the JT-3 became available, known as the JT-3D, the first aircraft to enter service with this engine doing so in March 1961. Turbofans had the advantage of being more efficient, more powerful and quieter.

As a result of research into problems encountered by the Comet, the Dash 80 was fitted with a flap on the leading edge of the wing. This significantly improved the stall margin during take-off and so was adapted as standard for the 707 and the KC-135.

Innovations with the 707 included seats mounted in rails, for easy fitment and removal. This made changing the seating configuration a simple operation. It also featured individual items called Passenger Service Units (PSUs). These were fitted above the seats and placed all the items required by a passenger in one compact unit. The PSU contained such items as reading lights, cabin crew call lights and oxygen masks. Another revolutionary feature was the plug-type door in the main cabin. This was a clever design that made the door larger than its hole in the fuselage (hence the term plug), but allowed it to swing outside the aircraft so that it didn't obstruct entry, and more importantly egress, from the cabin.

Having been well ahead of its rival Boeing made a few mistakes with its early 707s which cost it dearly. In trying to please its customers it offered several different versions, all of which ate up large amounts of money to develop. They also, under pressure from Pan Am who wanted more range and the ability to carry more payload, ended up redesigning the wing. However, that ultimately paid dividends as the version with the improved wing, which became the 707-320 series, sold well for many years. The last 707 variant rolled off Boeing's production line in May 1991, some 37 years after the Dash 80, and whilst there are only a few left in commercial use, a large number of KC-135 tankers are still in use today, albeit much modified (including being re-engined with new CFM 56 turbofans). Some of these aircraft are well over 50 years old and still serve with the USAF and a few other air forces. There are also a number of other military 707s

still in service, with no immediate signs of replacement. The last few of these to be built, such as the Royal Air Force's Sentry Airborne Early Warning (AEW) aircraft and the United States Navy's E-6 Mercury TACAMOs (TAke Charge And Move Out) were built with the CFM engine. Total production of 707s ran to 1,010 aircraft (most of which were for commercial use) and 820 C-135s.

By comparison only 114 Comets were built and the last one was retired from commercial service in 1980. A military version, the Nimrod, was developed but only 51 were built with the last few remaining in service until 2011. Attempts to rebuild and re-engine a small number proved very costly and problematic and the project was eventually scrapped in 2011, having been many years behind schedule and many millions of pounds over budget.

ABOVE KC-135R, 63-7999, departs Nellis AFB, Nevada, on 22 July 2010. *(Chris Wood)*

BELOW Like the US Navy's E-6 Mercury airborne command and control aircraft, the E-3D Sentry is also based on the 707-320, but with CFM56 engines. The RAF uses the Sentry for airborne surveillance and command and control. This is AEW1, ZH101, seen in spring 2011. *(Chris Wood)*

Douglas managed to sell 556 DC-8s with the last one rolling off its production line in 1972. In the late '70s a number were also re-engined with CFM 56 engines, which has prolonged their life, and there are still a few operating today.

Having cornered the market in long range air travel, Boeing started to look at the domestic market and saw a need for a medium-range jet. The challenge was to design an aircraft that could cruise at jet speeds but take off and land from the relatively short runways at the country's regional airports. To achieve this, it gave its design, which became the 727, a highly swept wing for high-speed flight, and triple-slotted trailing edge flaps plus leading edge devices along the whole leading edge, for slow speed. To help it stop on the ground it also featured the speed brake function of the spoilers, wheel brakes and reverse thrust from its three Pratt & Whitney JT8D turbofan engines.

Boeing's next project, which became the 737, was a small regional airliner to rival the second jet built by Douglas, the DC-9, and it was being worked on by one Joe Sutter in the summer of 1965. His major contribution to the design was the positioning of the engines, which had required some 'outside the box' thinking.

Enter the 747

Having been involved with the development of the 707, by 1965 Pan Am was after something bigger. Juan Trippe, Chairman of Pan American World Airways (Pan Am), asked

LEFT **When the seeds of the 747's design were sown in the mid-sixties, Boeing, Douglas and Lockheed were also working on designs for a large transport aircraft for the USAF, dubbed the CX-HLS, or Cargo Experimental – Heavy Logistics System. Lockheed won the bid with their C-5A.** *(PRM Aviation)*

Bill Allen, President of Boeing, to look into designing an aircraft that was much larger than the 707. Trippe reputedly said to Allen, 'if you build it, I'll buy it', to which Allen reputedly replied, 'if you buy it, I'll build it!'

The aircraft being proposed was way larger than anything then in existence. However at the time Boeing, Douglas (still its main rival in the commercial aircraft market) and Lockheed were working on a design for a very large transport aircraft for the USAF, known as the CX – HLS (Cargo Experimental – Heavy Logistics System). An aircraft of this size was made possible by the high bypass ratio turbofan engines then being developed by General Electric and Pratt & Whitney. In October 1965 Lockheed won the contract and their design evolved into the C-5A Galaxy powered by four of GE's TF-39 engines.

Although the 707 was a good design, the one thing Boeing had not allowed for in the design was growth potential. To make it able to carry greater payloads further would have required a major redesign with a stretched fuselage (something Douglas was able to do with the DC-8) which in turn would have required longer landing gear to avoid hitting the tail on take-off, and a redesigned stronger wing to lift greater loads. These options were

deemed too expensive for the projected returns.

At the time Boeing was heavily involved with a number of other major projects apart from the CX-HLS. It had teams working on the 727, 737 and a design for a SuperSonic Transport (SST), which evolved into the 2707. Supersonic transports were seen as the future, and a large subsonic transport was seen as an interim stopgap aircraft for use until the SSTs were in service. Consequently Boeing had its most experienced people working on the SST and was committing a vast amount of resources towards it. As a result of this, in August 1965 a relatively young engineer working on the 737 project named Joe Sutter was asked to head up the design team for the new large subsonic transport aircraft.

As Sutter and his small team got to work, their first goal was to decide the basic outline of the aircraft. As it was expected to be a pre-SST stopgap aircraft, it was not reckoned to sell well, so one of the first decisions Sutter made was to design it from the outset for passenger and freighter use. They proposed several sizes of passenger aircraft, with 250, 300 and 350 seats, and asked their airline customers what they wanted. They were surprised when the results came back; they all wanted the

350-seater! (707s and DC-8s of the time had around 140 seats.) Pan Am was pressing for a double-deck aircraft, much like Boeing's earlier troublesome Stratocruiser. However, the double-deck option was causing a few problems, mainly with cargo loading for the freighter version and passenger evacuation from the upper deck for the passenger version. After trying numerous manifestations of a double-decker, none of which were satisfactory, a very wide single-deck aircraft was suggested. As Sutter looked at the drawings of the single-deck alternative, he realised that this was the answer. For ease of freight loading it was decided to fit the freighter version with a hinged nose, which resulted in the flight deck being placed above the main deck, with an aerodynamic fairing behind it, which in turn gave rise to the 747s distinctive hump. In a month Sutter and his team had just designed the world's first wide-body jet airliner. All he had to do now was sell

the idea to the Boeing Board and, far more importantly, to Pan Am!

Sutter's announcement to the Boeing Board that he was going for a wide single-deck design, against the wishes of its main customer, did not go down too well. However Sutter's decision was respected by the Board but someone had to break the news to Pan Am. In early 1966 one of the design team was despatched to Pan Am's headquarters in New York. The Pan Am people listened but did not comment, and it was agreed that Boeing would build mock-ups of the two versions for Pan Am to look at.

Shortly afterwards a high-level Pan Am delegation, led by Trippe, visited Seattle. In anticipation of this, before the New York meeting Boeing had actually already started building the two wooden mock-ups, one of a double-decker and one of the very wide single-decker. The Pan Am people were shown around the double-deck version first, and were shocked to see just how far the upper deck would be above the ground. They were then shown around the cavernous single main deck version. Boeing had also built a mock-up of the flight deck area and Trippe asked what the space behind would be used for. One of his team suggested it could be used as a crew rest area, but Trippe responded that it would be 'reserved for passengers'. At that point Sutter realised that Trippe had been convinced, and as they left the mock-up Trippe told him he had made the right decision. The 747 was officially launched in March 1966 and Pan Am signed a contract for 25 aircraft the following month.

As the design work continued, it was decided to leave room for development of the aircraft, a lesson learnt from the 707. The wing area was increased from the initial plan of 5,200sq ft to 5,500sq ft. The landing gear was designed to support weights far greater than were being planned initially. Boeing has a 'safe design' philosophy that means they design their aircraft to be as safe as possible, and also easy to fly in airline service. This means they design their aircraft with back-up systems, referred to as redundancy, to cater for systems failures. With the 747 Sutter decided to go for four systems to give plenty of redundancy. It was designed with four engines, four electrical channels, four hydraulic systems and four main landing gear legs. He received some criticism for this but his decision was vindicated in the early days of 747 operations when an incident with a Pan Am aircraft taking off from San Francisco resulted in it losing two of its main gear legs and three of its hydraulic systems. The crew were able to maintain control of the aircraft and bring it around for a safe landing.

One of the deciding factors for the width of the fuselage had been the ability to load two standard 8ft by 8ft freight containers side by side. It wasn't intended that the containers would fly, but it was felt that by having the dimensions the same, it would be easy to load cargo on and off from and into containers. The wings had to be swept for high-speed flight, but also needed to be capable of low-speed flight for take-off and landing. The 707 had wings swept to 35° and cruised at a speed around M.80 (Mach.80 = 80% of the speed of sound). Pan Am originally wanted the 747 to cruise at M.90, which would have required a wing sweep of 40°. Eventually it was decided to compromise and give the 747 a wing sweep of 37.5° and a cruise speed around M.85. To achieve the low-speed capabilities, the aircraft was fitted with triple-slotted trailing edge flaps and leading edge flaps (earlier Boeing aircraft had leading edge slats).

The weights being planned for the aircraft resulted in it being fitted with four main landing gear legs, two fitted to the fuselage and two to the wings, each with four wheels. This reduced the weight distribution on the runways and taxiways and also gave the aircraft excellent braking capability, with 16 main wheels each with its own brake.

Another major hurdle for the 747 programme was that due to its size, Boeing did not have anywhere big enough to build it! They had to construct a whole new factory for this purpose, and having considered several locations, decided on a site at Everett, north of Seattle. The building became the world's largest, as attested to by the *Guinness Book of World Records*, and was still under construction when 747s started rolling off the production line.

By the time the prototype 747 was rolled out of the Everett factory on 30 September 1968, 26 airlines had signed orders for it. However it was nowhere near ready to fly. It had been hoped to make the first flight on 17 December 1968, to mark the 65th Anniversary of the Wright brothers' first flight at Kitty Hawk, South Carolina. However, that was not to be, but it did eventually take to the air for the first time on 9 February 1969, flown by Jack Waddell, Brien Wygle and Flight Engineer Jess Wallick. The

LEFT With the great increase in weight over its predecessor the 707, it was necessary to fit a 16-wheel 4-leg main undercarriage to the 747. *(Chris Wood)*

RIGHT The 747 prototype under construction at Everett in 1968. *(PRM Aviation)*

BELOW Great ceremony hailed the roll-out of the 747 prototype on 30 September 1968. *(PRM Aviation)*

prototype aircraft then started a rigorous flight-test phase, along with the first four production models.

Various problems were encountered during this phase, the most significant being with the engines. Pratt & Whitney were having a lot of problems with the JT9D, they were not as efficient as they had hoped, or as reliable, and they were prone to surging. At one stage production of the 747 outstripped Pratt & Whitney's ability to supply enough engines, leaving dozens of 747s sitting outside the Everett factory awaiting their powerplants. However, with help from Boeing, the problems were eventually resolved and the 747 was awarded its Approved Type Certificate by the Federal Aviation Authority (FAA) on 30 December 1969. Shortly afterwards Pan Am received their first aircraft and *Clipper Victor* performed the first commercial service with a flight leaving in the early hours of 22 January 1970 from New York's John F. Kennedy Airport, arriving into London's Heathrow Airport later that day. By 16 July the millionth passenger had flown in the 747.

ABOVE On 9 February 1969 the prototype took to the air for the first time from Paine Field, Everett, with test pilot Jack Waddell and co-pilot Brien Wygle at the controls. *(PRM Aviation)*

BELOW Delays in certification of the Pratt & Whitney JT9D engine meant that by early 1970 some 30 engineless 747s were parked outside the Everett plant with concrete blocks hanging from their engine mounts. When finally delivered, the JT9D became the first high bypass ratio jet to power a wide-body aircraft. *(PRM Aviation)*

London début for the 747

In the run-up to US FAA approval, Boeing had begun delivering 747s to Pan Am in 1969. Ahead of the type's service debut, set for 21 January 1970, Pan Am flew a proving flight with N735PA, *Clipper Constitution,* between New York and London on 12 January as part of a planned European tour.

Clipper Constitution disgorged its 362 passengers with little difficulty and, apart from the loss of a few items of baggage, the temporary handling arrangements at Heathrow worked well. The first passengers were boarding buses and all had cleared

Customs within half an hour of the aircraft arriving at the stand.

Engines were started for the demonstration flight at about 2:15pm, but No 1 failed to light. A lack of fuel pressure as the JT9D reached light-up rotation speed was found to be the cause. Several attempted starts produced nothing more than a hot starter motor, but all four engines were eventually fired up at 4:05pm. Further trouble with the JT9D engines and bad weather resulted in the cancellation of the rest of the five-capital European tour, and the 747 departed the next day, some 23 hours late, heading for New York. The inaugural Pan Am commercial flight of what the press had dubbed the 'jumbo jet', due to be operated on 21 January 1970, was postponed owing to technical problems when the original aircraft allocated to Flight PA2, *Clipper Young America* (N735PA, renamed), suffered engine problems and had to return to the gate. *Clipper Victor* (N736PA) stepped in and finally left New York–John F. Kennedy Airport at 1:52am on 22 January, arriving at London–Heathrow at 2:02pm after a 6hr 10min flight. (*PA Photos*)

RIGHT Less than six months into commercial service, Boeing 747s had carried their millionth passenger. No other airliner had reached this milestone so quickly. The reason was, of course, that the 747 was attracting load factors far above industry averages. Airlines reported traffic increases of up to 78.8% on their 747 routes. This is the scene at London–Heathrow in 1971 with 747s from BOAC, Pan Am and TWA on the ramp. (*Jonathan Falconer*)

FLYING THE BOEING 747-100

Captain Henry Chambers

Henry Chambers spent 29 years as a pilot with BOAC and British Airways from 1957 to 1986. He first became involved with the Boeing 747 in 1976. In the early '50s he had spent six years in the RAF where he flew the first jet fighters – Vampire, Venom and Meteor. He then joined BOAC and spent seven years on the Britannia and eleven years on the VC10 before bidding for British Airways' 747 fleet.

The training programme for the 747 differed from that used on earlier types, being quicker and concentrating on the simulator to such an extent that actual flying training could be kept to a minimum. It began with four weeks in the classroom followed by about 40 hours in the simulator, before having 8 hours of instructional flying. Finally four trips over the Atlantic under supervision were completed for full pilot certification. The simulator was equipped with three-axis movement and vastly improved visuals. It also flew very like the actual aircraft.

Although the VC10 was a wonderful aircraft to fly, the 747 was even better. Simply stepping into the flight deck and sliding into that comfortable seat was a pleasure. I suppose it was a bit like getting into a Cadillac after being used to a Cortina. The size was very soon forgotten and it was surprisingly responsive and manoeuvrable. Even the feeling of landing from what felt like the top of a double-decker bus presented no problem. It is hard to compare the difference in cockpit workloads between the VC10 and the 747 because although there was more automation on the 747 everything still had to be closely monitored. The pilot flies his aircraft according to the indications on the instrument panel in front of him and all the rest follows on behind – and it really doesn't matter if there are four passengers sitting back there or 400.

The 747 was, of course, the first wide-bodied aircraft in service, the first to have an upper deck and the first to be powered by a bypass fan engine. The Pratt & Whitney JT9D-3 was used on the first aircraft and was sadly under-powered. For instance, after take-off the aircraft had to be levelled off in order to build up sufficient speed to retract the flaps. But this was soon replaced by the JT9D-7, which was much better. However, it too had an odd quirk. We all like to handle the throttles gently and smoothly, but if this was done at top of descent a flame-out of one or more engines could follow. So the flight engineer was obliged to retard the throttles to idle power very firmly and quickly. A simple answer but it worked.

Navigation was achieved by the INS (Inertial Navigation System). This consisted of a computer and several sophisticated gyroscopes, which monitored every slight movement of the aircraft. The required route could be fed into the system, a few waypoints at a time, and in flight this would be followed by the autopilot. It was very accurate and made all other long-range navigation aids redundant, including the navigator. The GPS was to come along much later.

The 747-200 arrived quite soon afterwards. With its more powerful engines, the Rolls-Royce RB211 in BA's case, it had a larger load capacity and greater range making further airframe development possible (cargo, Combi and stretched upper-deck versions). The 200 fitted seamlessly into our operations with smooth and reliable powerplants. None of these versions required further pilot training – handling

RIGHT Captain Henry Chambers.
(Henry Chambers)

was the same although their capabilities differed, but the 400 version was another story.

There were hardly any other problems to bother us so that very soon we were confident in the aircraft and enjoyed operating it. With its dual autopilot, autothrottle and auto-braking it was well equipped to carry out automatic landings in almost nil visibility.

Because it was so much larger than any other aircraft in use there were plenty of headaches for the airport authorities. Larger parking spaces were needed; passenger jetties required modification; tugs for manoeuvring the B747 had to be much more powerful and taxiways had to be surveyed to remove obstructions because of the large wingspan. Pilots needed to be confident that if they followed the taxiway centre lines they would not touch anything. Airport authorities responded with enthusiasm and quickly brought their airports up to the required standards – they were obviously keen to host this iconic aircraft.

The passenger capacity was much greater than with any previous commercial aircraft but an added bonus, due to its wide body, was the much bigger hold that enabled large amounts of freight to be carried as well as passenger baggage. From Nairobi it was possible to take a ton or two of pineapples home to London, whereas the VC10 had been much more limited when it came to freight.

The routes we flew initially were mainly across the Atlantic but very soon we were operating to the Far East, South Africa and the Caribbean, and just before I retired I was able to operate one trip to the Falklands. One route that I always enjoyed was London to Japan via Anchorage. This took us just a few hundred miles from the North Pole and, more often than not, we had the pleasure of a lovely display of the Northern Lights. The stopover in Anchorage also gave me the opportunity to get back on skis after an absence of over 25 years. This was most enjoyable but it was also in Anchorage that I had an experience I would not have wanted to repeat. As we neared the end of our landing run the aircraft very slowly veered to the left of the ice-covered runway and came to a stop in 2ft of snow. There was a very strong crosswind blowing and this contributed to the fact that two of the engines shut themselves down as reverse power was selected. There was no damage and the only injury was to my pride.

Flying the B747 was both a pleasure and a privilege and most of us were impatient to get started. I suppose there was some kudos in being a 'jumbo' pilot but this came mainly from outside the airline. Within it many people were just jealous or claimed they would rather be on Concorde. I am proud to have played a small part in its history.

ABOVE The 747-200B was an improvement over the -100, with increased fuel capacity and more powerful engines. This is BA's -211B, G-GLYN, at Gatwick in July 1990.
(Chris Wood)

LEFT **The longer-range 747-200 entered airline service in January 1971. This is HS-TGB at Paris–Charles de Gaulle on 29 December 1991.** *(Chris Wood)*

CENTRE **There were no all-cargo versions of the 747-100 built at the time, but with the advent of the 200-series in 1971 its improved weights made it ideal as a dedicated freighter. Here is Flying Tigers' 747-245F, N816FT.** *(Jonathan Falconer)*

Before the first aircraft had actually taken to the air, the airlines were pressing Boeing for a longer-range version. In November 1968, Boeing announced the launch of the 747-200 series, with the first aircraft being rolled out on 10 September 1970. It first flew on 11 October and the first one was delivered to KLM of the Netherlands on 15 January 1971.

Although having originally designed it to be a freighter, the -100 was not built in a freighter version but the improved weights of the -200 made this viable and the first -200 freighter was rolled out on 23 November 1971. It made its first flight on 30 November and entered service with Lufthansa of Germany on 19 April 1972.

Various other versions of the standard 747 were produced, including the 747-100SR (Short Range) for the Japanese domestic market and two that could carry a mix of passengers or freight, the -200C (Convertible) and the -200M (Combi). The -200C featured the nose cargo door configuration, or a mix of both. The -200M featured the side cargo door and the options were an all-passenger configuration or a mix of both passenger and cargo. These versions were aimed at airlines that did not have the requirement for a full passenger 747, or seasonal changes in payload.

LEFT **Designed primarily for Pan Am's West Coast to Japan route, the 747SP – or Special Performance – had a shorter fuselage than a standard 747. It was fitted with a taller tail plane to compensate for the consequent handling differences, thereby giving it the flying characteristics as its full-size cousins. This is Air China's B-2454, seen at London–Gatwick, 21 May 1992.** *(Chris Wood)*

RIGHT South African Airways' B747-300, ZS-SAU, *Kaapstad*, pictured at London–Heathrow inside British Airways' maintenance area at Hatton Cross in 1990. *(Jonathan Falconer)*

The most unusual version was the short-bodied 747SP (Special Performance). Making the fuselage smaller but keeping the wings the same gave this version a greater range. The SP's shorter fuselage meant it needed a bigger rudder, so it was provided with a taller tail fin, and to keep its weight down it was fitted with single-stage flaps. However, an important design requirement was that it should handle like the -100 and -200, so that pilots would have no problems flying all versions.

Pan Am had acquired the long-range 'Special Performance' jumbo to enable the airline to operate a non-stop service between the United States and Japan and vice-versa, supporting its round-the-world routing. On 25 April 1976 Pan Am inaugurated the Los Angeles/New York to Tokyo flight service. Only 45 examples of the SP were built but, like the Combi version, it introduced the 747 to airlines that were not in a position to buy a 'full-size' one at the time. Subsequently, many of them did.

A version with a Stretched Upper Deck (SUD) known as the -300 appeared in 1983. This had the same maximum take-off weight as the later -200s but the basic aircraft was heavier, so whilst it could carry more passengers than the -200 it did not have as much range.

All of these versions of the 747 have come to be known as the '747 Classic'.

BOEING 747 WORLD RECORDS

On 1 May 1976 Pan Am Boeing 747SP, N539PA, *Clipper Liberty Bell*, flew from New York round the world via Delhi and Tokyo, returning to New York on 3 May. Pan Am had offered a special invitation to 130 fare-paying passengers to fly on this specially scheduled flight. The flight of *Clipper Liberty Bell* set a new round-the-world flight speed record of 46 hours, 0 minutes and 50 seconds for its 23,137-mile circumnavigation of the earth.

On 28–30 October 1977, Pan Am Boeing 747SP, N533PA, *Clipper Freedom*, set a new round-the-world record to celebrate the 50th anniversary of Pan Am. Flight 50 was flown from San Francisco over a distance of 26,706 miles in 54 hours, 7 minutes and 12 seconds. Three stopovers were made at London–Heathrow, Cape Town, and Auckland. Flight 50 flew over both poles.

BELOW All of Singapore Airlines' Boeing 747-300 'Big Tops' were powered by Pratt & Whitney JT9D-7R4G2 engines. 9V-SKA, illustrated, was the first of 14 747-312s for the carrier. Delivered in April 1983 it served with Singapore until its retirement in 1999. *(Jonathan Falconer)*

SPECIAL USE 747S

A number of Classic 747s were built for roles that are somewhat different to the type's original purpose. Some are in use as VIP transports of which the best known are the pair of -200s operated by the USAF for the President of the United States (designated as the VC-25) and known as Air Force One when the President is on board. The USAF also operates a further four -200s as the E-4 Advanced Airborne Command Post. Other proposals have been for a 747 cruise missile launch platform and another to launch parasite fighter aircraft!

Several 747s have been converted to other more unusual roles. NASA used two heavily modified -100s to transport its space shuttles and Evergreen International converted a pair (a -100 and a -200) into Supertankers for firefighting.

The SP model has proved popular in the VIP role for which a number have been converted. One SP aircraft has also been modified to carry a 2.5m-diameter infrared telescope, weighing over 17,000kg. The aircraft is operated by NASA as an airborne observatory and is known as the Stratospheric Observatory for Infrared Astronomy (SOFIA).

The trend for conversion of the 747 continues with the -400, examples of which can be seen elsewhere in this book.

BELOW President Barack Obama meets his staff in the Air Force One conference room, en route from London–Stansted to Strasbourg, 3 April 2009. *(The White House, Washington)*

LEFT No matter where in the world the President of the United States travels, if he flies in a USAF jet the aircraft is called 'Air Force One'. Technically, 'Air Force One' is the call sign of any Air Force aircraft carrying the President. In practice, however, Air Force One is used to refer to one of two highly customised Boeing 747-200B series aircraft. The USAF designation for the aircraft is VC-25A.

Capable of being refuelled in mid-air, Air Force One has unlimited range and can carry the President wherever he needs to travel. The onboard electronics are hardened to protect against an electromagnetic pulse, and Air Force One is equipped with advanced secure communications equipment that allow the aircraft to function as a mobile command centre in the event of an attack on the United States.

Inside, the President and his travel companions enjoy 4,000sq ft of floor space on three levels, including an extensive suite for the President that features a large office, lavatory, and conference room. Air Force One includes a medical suite that can function as an operating theatre, and a doctor is permanently on board. The aircraft's two food preparation galleys can feed 100 people at a time.

Air Force One also has quarters for those who accompany the President, including senior advisers, Secret Service officers, travelling press, and other guests. Several cargo planes typically fly ahead of Air Force One to provide the President with services needed in remote locations. This is Air Force One, 82-28000, leaving Las Vegas on 26 January 2012. *(Chris Wood)*

RIGHT Perhaps the most distinctive and immediately recognisable specialist conversion is the B747-Shuttle combo. NASA used two modified Boeing 747s as Space Shuttle Carrier Aircraft (SCA). One is a 747-100, the other a 747-100SR (Short Range). The two aircraft are identical in appearance and in their performance as Shuttle Carrier Aircraft. The SCAs were used to ferry space shuttle orbiters from landing sites back to the launch complex at the Kennedy Space Center and also to and from other locations too distant for the orbiters to be delivered by ground transportation. The orbiters are placed atop the SCAs by mate-demate devices, large gantry-like structures that hoist the orbiters off the ground for post-flight servicing and then mate them with the SCAs for ferry flights. Here, shuttle *Columbia* hitches a ride on SCA 905 for the flight from Palmdale, California, to Kennedy Space Center, Florida, on 1 March 2001. *(NASA)*

RIGHT Work is under way on the Shuttle Landing Facility at Kennedy Space Center to separate space shuttle *Atlantis* from the SCA via the mate/demate device. *(NASA)*

BELOW A hoist is attached to *Atlantis*, which suspends the shuttle while the SCA is moved away from underneath it. *(NASA)*

BELOW NASA has transferred space shuttle *Discovery* to the US National Air and Space Museum in Washington to begin its new mission to commemorate past achievements in space and to educate and inspire future generations of explorers. Here *Discovery* is pictured mounted atop Boeing 747 SCA 905 as it departs on its final flight to Washington DC on 17 April 2012. *(Christopher Parypa/Shutterstock.com)*

ABOVE Evergreen International Aviation was the first company to convert a Boeing 747 into an aerial firefighting supertanker aircraft. Boeing 747-100, N479EV, was originally built in 1970 for Delta Airlines. With a capacity of 20,500 US gal (77,600 litres), the 747 Supertanker is the largest aerial fire-fighting aircraft in the world. Two such 747s have now been converted to the role. *(Evergreen International Aviation, Inc.)*

BELOW Even with 20,500gal of retardant on board, the Supertanker is still 150,000lb below its maximum take-off weight. *(Evergreen International Aviation, Inc.)*

ABOVE AND LEFT The aircraft's tank system allows it to make segmented drops, which means that it can release its 20,500gal either in one continuous 'overwhelming response', at multiple intervals while in flight, or at a speed equivalent to falling rain. *(Evergreen International Aviation, Inc.)*

BELOW The Supertanker's retardant dropping speed is approximately 140kt, which provides a 30 per cent cushion over the Boeing 747's stall speed. During a retardant drop the aircraft is configured as if it were flying a final approach for a landing, at a safe altitude of between 300 and 600ft. *(Evergreen International Aviation, Inc.)*

BELOW Evergreen's Supertanker disperses pressurised retardant from four nozzles in the fuselage underside. The system enables the aircraft to fight fires from higher altitudes than is normally the case with aircraft that use a gravity-drop system. *(Evergreen International Aviation, Inc.)*

RIGHT Dubbed by some commentators as the 'flying light-sabre', the YAL-1 Airborne Laser Test Bed (ALTB) is an advanced platform for the US Department of Defense's directed energy research programme. Using two solid state lasers and a megawatt-class Chemical Oxygen Iodine Laser (COIL) housed aboard a modified Boeing 747-400 Freighter, the ALTB demonstrates the potential of using directed energy as a viable technology against ballistic missiles. The project has now been cancelled. *(USAF)*

RIGHT The Government of Japan maintains two Boeing 747-400 aircraft to transport the Prime Minister and other officials on overseas visits. With the radio call signs 'Japanese Air Force One' and 'Japanese Air Force Two', each aircraft can carry up to 140 passengers and when required they can be used for the emergency evacuation of Japanese citizens and overseas deployment of Japan Self Defense Forces personnel. The aircraft always fly together on Government missions, with one serving as the primary transport and the other as a back-up with maintenance personnel on board. This is 747-400, 20-1102, departing London–Heathrow on 7 March 2010. *(Tony Osborne)*

RIGHT Preserving a 747 is not an easy task, given its size, but a few have tried and examples of the Classic 747 are to be found at places such as the Aviodome at Lelystad in the Netherlands, and at the Technik Museum at Speyer in Germany. The very first 747 is on display at the Museum of Flight at Boeing Field in Seattle.

Turning an unwanted aeroplane into a restaurant is not a new idea; they can be found all over the world, but it's not often that you find one as big as a 747! Pan Am's first aircraft was turned into a restaurant in South Korea, but it has now been scrapped.

In Sweden an enterprising individual went further and turned a 747-200 into a hotel. The Jumbo Stay Hostel (pictured) opened in 2009 and can be found outside Stockholm's Arlanda Airport. It features a bar, restaurant and 29 rooms, with 76 beds between them. *(Jonny Kristoffersson/iStock)*

ABOVE Boeing 747-400s nearing completion on the Everett production line. *(PRM Aviation)*

BELOW B747-400 prototype in flight. *(PRM Aviation)*

BELOW Singapore Airlines' (SIA) first 747-400 (known as the 'MEGATOP') arrived in 1989. SIA became the largest operator of the 747-400, with a fleet of 59 aircraft. Their last passenger aircraft was withdrawn in 2012, but SIA Cargo still operates 13 747-400 Freighters. 747-400, 9V SMJ, is pictured at Paris–Charles de Gaulle on 29 December 1991. *(Chris Wood)*

The 747-400

By the early 1980s orders for 747s were dwindling. However, at the same time the economies of countries around the Pacific Rim were booming, opening up new longer-range routes that were not economically viable at that time with existing versions of the 747. Although Boeing was already working on a long-range version of the 747-300, known as the 747-300A (300 Advanced), a group of their major airline customers got together to put pressure on the manufacturer to go further and completely redesign the aircraft to take advantage of new technologies that were then becoming available.

The first major redesign of the 747 was announced in October 1985 with the -400 version. This featured some major changes – an improved and longer wing (built from stronger and lighter aluminium alloys) fitted with winglets, a redesigned wing to fuselage fairing, improved engines (which, together, made it significantly more efficient than the 'Classic' versions, burning roughly 15% less fuel), more rudder travel, carbon wheel brakes, the stretched upper deck of the -300 and the 'glass cockpit' that had been designed for Boeing's 757 and 767 aircraft. Coupled with a large increase in automation this allowed a reduction of the cockpit crew to just two pilots.

There were also a host of other refinements to make the aircraft easier to operate, not just for the pilots but also for the engineers. The entire aeroplane is connected to a Centralised Maintenance Computer System (CMCS) which continually monitors all the systems and detects any faults. The engineers can download this information and see straight away where the problems are, saving considerable amounts of time. They can even access the CMCS whilst the aircraft is airborne, which means that spare parts, equipment and maintenance personnel can be ready when the aircraft lands to immediately carry out any repairs, rather than first having to spend time identifying the problem.

The aircraft was offered with a choice of three models of engine from Pratt & Whitney (P&W), General Electric (GE) and Rolls-Royce (RR). The first aircraft, powered by P&W engines, was also rolled out at Everett on 26 January 1988, on the same day as the prototype of Boeing's 737-400 was also rolled out at Renton. It first flew on 29 April 1988 and entered service with Northwest Airlines on 9 February 1989 (the 20th anniversary of the 747's first flight). The first GE-powered aircraft flew on 27 June 1988 and this was eventually delivered to Germany's Lufthansa. The first RR-powered aircraft made its first flight on 28 August 1988, and this was delivered to Hong Kong's Cathay Pacific.

Just like the 'Classic' versions of the 747, Boeing designed a range of 747-400 variants. After the standard passenger version, next was the -400M Combi. It features a side cargo door, a strengthened floor in the rear cargo section of the fuselage, a cargo handling system, fuselage strengthening and an increased Maximum Zero Fuel Weight. The first one was rolled out on 23 March 1989, made its maiden flight on 30 June and entered service with KLM on 12 September.

Next came the -400D (Domestic). Like the earlier 100SR this model was for the Japanese domestic market. The 400D features a strengthened upper-deck floor, an extra three upper-deck windows on either side plus a range of additional structural strengthening. Its most distinguishing feature is the absence of the winglets, and it also does not have the wingtip extension of the standard 400. It has the same higher maximum zero fuel weight as the 400 Combi, but a much lower maximum take-off weight than the other variants. The first example was rolled out on 18 February 1991, first flew

on 18 March and entered service with Japan Airlines on 22 October.

Boeing designed the 400D with the ability for it to be converted into a standard 400. This requires the addition of the wingtip extensions and winglets, and once achieved the modified aircraft has the same maximum take-off weight as the standard aircraft. However it is a one-way operation – it cannot be converted back to a 400D.

A freighter version, the -400F followed in 1993. Its most distinctive feature is the lack of the stretched upper deck. It has a hinged main deck nose cargo door, an optional side cargo

ABOVE The most distinguishing features of the 747-400D (Domestic) are the absence of winglets and wingtip extensions of the standard 400. This is All Nippon's 400D, JA8961, at Tokyo–Haneda International Airport (which handles almost all domestic flights to and from Tokyo) on 5 April 1994. (Chris Wood)

LEFT A hinged main-deck cargo door in the nose is the principal feature of the 747-400F. (iStock)

door (like the Combi) and a main-deck cargo handling system. The first one was rolled out on 8 March 1993, first flew on 4 May and entered service with Cargolux of Luxembourg on 17 November.

Further refinements occurred in 2000 with the launch of the 400ER (Extended Range) and in 2001 with the 400ERF (Extended Range Freighter). These featured additional fuel tanks and increased weights. The first 400ER was rolled out on 17 June 2002, first flew on 31 July and entered service with Qantas of Australia on 31 October. It was pipped by the 400ERF,

which entered service with Air France on 17 October.

In the early part of the 21st century orders for the passenger version of the 400 dwindled and the last example of this series, which became the 1,357th 747 to be built, made its first flight on 8 April 2005. It was subsequently delivered to China Airlines of Taiwan. There were still orders for freighters, both for the 400F and the 400ERF, and production continued with the last aircraft being delivered in 2009. In total Boeing built 694 747-400s, compared to 724 747 Classics. However, that is not the end of the story.

(Far more detail about the road that led to the design and construction of the Boeing 747 can be found in the excellent *Wide-Body* by Clive Irving, and also in Joe Sutter's fascinating autobiography *747, Creating the World's first Jumbo Jet* and other adventures from a life in aviation.)

Beyond the 747-400

Several developments of the 747-400 have been proposed since its launch, none of which have come to fruition. The first was in 1996 when Boeing announced plans for the 747-500X and -600X versions, officially launching them at that year's Farnborough Air Show. Both models used the basic -400 fuselage but stretched it. They were to feature a redesigned wing and more powerful engines, the Engine Alliance (a joint venture between General Electric and Pratt & Whitney) GP 7176 or the Rolls-Royce Trent 900. They were also to incorporate a lot of the technology developed for the 777, including a fly-by-wire system. The -500X had an extra 5.5m (18ft) of fuselage and a maximum take-off weight of 450 tons and was expected to carry slightly more than the -400 but over a longer range. The -600X added another 8.8m (29ft) to the fuselage and had a maximum take-off weight of 540 tons and was expected to carry significantly more (around 30%) than the -400 but over a similar range. To cope with these increased weights they would have additional wheels, with 20 main wheels and 4 nose wheels.

A further design study of this time, the -700X, featured a widened fuselage with the wing of the -500X and -600X. This was intended as a counter to the large aircraft being proposed by Airbus, its A3XX. The prohibitive cost (quoted at around US $5 billion in 1996) of developing these versions and the lack of sufficient orders meant that they didn't get further than the drawing board. Meanwhile, the A3XX developed into the A380, which finally entered service in October 2007, several years behind schedule and several billion dollars over budget.

In 2000 Boeing unveiled further developments called the 747-400X, the 747X and 747X Stretch. Freighter and Combi versions were also under consideration. The -400X was effectively a tweak of the existing -400, which gave it additional fuel tanks and a higher maximum take-off weight. The 747X and X Stretch would feature a wing root extension, increasing the span to 69.8m (229ft), and were to be powered by either the Engine Alliance GP7172 or the Rolls-Royce Trent 600. The 747X and X Stretch also did not generate enough interest to justify progression beyond the drawing board, but the 747-400X did, later becoming the 747-400ER.

Boeing's next offering was announced at the Asian Aerospace Show in Singapore in February 2002. This was a version known as the 747-400XQLR (Quiet Long Range). Again this was offered in both passenger and freight versions. It was to feature modifications to improve efficiency and reduce noise, such as redesigned wingtips (like the 777) rather than winglets and saw-tooth engine nacelles to reduce noise. This version also failed to get past the design stage but many of its features found their way on to Boeing's next offering, the 747 Advanced, which became the 747-8.

In late 2002 it was reported that Boeing came up with another proposal, known as the 747-800X. This drew on changes considered for the 400LRX and added a 2m fuselage stretch in front of the wing, plus additional fuel capacity in the horizontal stabiliser.

BELOW In 1996 Boeing announced the 747-500X and -600X at the Farnborough Air Show. The cost of the changes from previous 747 models, in particular a new 251ft-span wing, was estimated at more than US$5 billion. Boeing cancelled the project when it was not able to attract enough customer interest. *(Boeing)*

Into the future – the 747-8

First unveiled in June 2003 as the 747
Advanced, this version was formally
launched on 14 November 2005 as the 747-
8 family. Initially there are two versions, the 747-
8F (Freighter) and the 747-8I (Intercontinental).
The launch customers were Cargolux of
Luxembourg (who had also been launch
customer for the -400F) and Nippon Cargo
of Japan for the 8F, and Lufthansa for the 8I.
It was a major redesign of the 747 and drew
on technology from Boeing's 787 (hence the
designation 747-*8*), but at the same time it
has retained many features of the -400. It
also benefits from features used in some of
Boeing's other aircraft, such as their highly

successful 777, later versions of the 767 and
even the 747SP.

One of Boeing's design goals for the -8
was commonality with the -400. The objective
was that pilots would be able to fly both
variants with minimal additional training; that
the same ground servicing equipment could
be used for both versions; and that airports
supporting operations by 747-400s could
also support the 747-8 without needing any
upgrading. It is the same height as the -400
so the same airport ramp equipment as the
-400 can be used. One exception to this is
that, due to its higher weights, it requires a
bigger tow bar and it has larger nose wheel
tyres, so some tow bar-less tow vehicles may
be unsuitable.

The major changes embodied in the -8
were a longer fuselage, a new supercritical

wing and new engines. It is built primarily of new generation aluminium alloys, but features graphite composite flaps, spoilers, rudders and engine nacelles. These are more durable and lighter than the materials used on the -400, thereby reducing weight even further and making the aircraft more efficient.

For the first time in the 747's history the fuselage has been stretched, with a 4.1m (12ft 4in) plug inserted in front of the wing and a 1.5m (5ft) plug behind. As a result of the longer fuselage the aircraft has been fitted with an electronic tail-strike protection system.

The new wing has a higher aspect ratio than the wing on the -400, giving it improved aerodynamics and it is actually thicker so its fuel capacity is greater, at around 194 tons (at a specific gravity of 0.8). It features a raked wingtip instead of the winglet of the -400. The inboard trailing edge flaps are double-slotted, the outboard are single-slotted (which reduces weight), and the ailerons droop. The wing leading edge features improved Krueger flaps. The spoilers and the outboard ailerons are fly-by-wire, which reduces weight but also allows an element of computer control into the flight controls. The aerodynamic improvements mean that the -8's approach speed at Maximum Landing Weight (MLW) is not much higher than the -400 at MLW, even though the -8 has a significantly higher MLW.

There is only one engine option for the -8, the General Electric GEnx-2B67, which produces 67,500lb of thrust. It is very similar to the engine fitted to the Boeing 787 but features a smaller fan and, unlike the 787's engine, it has an air bleed capability. The single-stage fan has composite blades with titanium leading edges. This is followed by a 3-stage LP compressor, a 10-stage HP compressor, a 2-stage HP turbine and a 6-stage LP turbine. It features a low emission twin annular premixing swirler (TAPS) combuster, which helps to reduce emissions. Like the engines for the -400, the engine is modular, and is split into two sections, the fan and the core. Compared to the CF6-80 this engine has five fewer stages and 24% fewer parts.

To aid noise reduction the inside of the engine nacelle is fitted with a sound-absorbing liner, and there are external nacelle chevrons at the rear and on the exhaust nozzle. There are other refinements such as much smaller wing anti-ice exhaust ports, which overall achieve a 25% reduction in noise when compared to the -400.

The -8 has a new environmental control system with digital control, which is lighter and more reliable.

The flight deck includes some refinements, such as provision for an Electronic Flight Bag (EFB), electronic checklists, an Airport Moving

BELOW Powered by four General Electric GEnx-2B67 engines, on 20 March 2011 the 747-8I (Intercontinental) made its maiden flight from Paine Field, Washington State, to nearby Boeing Field in Seattle, where the company's flight-test operations are housed. *(Chris Wood)*

Map (AMM) display and a GPS-based Global Landing System (GLS).

Despite all the changes and refinements, Boeing states that a current -400 pilot will only need three days of classroom instruction to be ready to fly the -8.

Another improvement is the time between major services has been increased significantly over those for the -400: line maintenance: 1,000 flying hours; hangar maintenance: 10,000 flying hours or 24 months; heavy maintenance: 8, 8 then 6 years.

The first flight of the -8F took place on 8 February 2010, one day before the 41st anniversary of the first flight of the very first 747. The first flight of the -8I followed just over a year later, on 20 March 2011. The first -8 to enter service did so on 12 October 2011 with

Cargolux after a very short delivery flight from Everett to Seattle–Tacoma International Airport, where it went straight into revenue service.

The competition

The 747 was the world's first wide-bodied jet airliner. Following not far behind it were offerings from McDonnell Douglas in the shape of their DC-10 and from Lockheed with their L1011 TriStar. Both of these aircraft were tri-jets and were initially designed for the US domestic market, which meant they had the range to cross the United States but no further. Later developments of both designs saw longer-range models introduced, and while both sold reasonably well (446 DC-10s were built and 250 TriStars) they failed to achieve anywhere near the sales that Boeing did with the 747. When the last 747-400 rolled off the production line Boeing had built 1,418 747s (plus the prototype), and production continues with the -8 variant.

McDonnell Douglas redesigned the DC-10 and produced the MD-11, in much the same way that Boeing had redesigned the 747 with the -400 version. However, this did not sell as well as expected with only 200 being built, and production ceased in 2000. Most MD-11s and the few remaining DC-10s that are still flying have since been converted to freighters. McDonnell Douglas was taken over by Boeing in 1997.

The TriStar was Lockheed's final attempt at an airliner and only a very small number remain in service today.

Across the Atlantic in Europe in the early 1970s a consortium was formed to develop a new airliner for the European market. The aircraft was dubbed the Airbus A300 and was planned to seat 300 passengers, but only had enough range to suit the major city pairings in Europe. Later versions had increased range and a total of 561 were built. However, this aircraft's most important legacy is the family of jet airliners it spawned. Today, Airbus is Boeing's only serious rival in the airliner manufacturing business.

In January 1986 Airbus announced the launch of a four-engine long-range aircraft, the

BUYING A 747

Aeroplanes of this size are not cheap and very few airlines or individuals can afford to buy them outright. When Pan Am ordered their first 747s they paid US$525 million for 25 aircraft (US$21 million per aircraft). When the -400 started to roll off the production line in 1988, the basic price was around US$120 million an aircraft and by 2008 the list price had risen to between US$234 to US$268 million. For the -8 it ranges from US$293 to US$308 million.

Financing the purchase of one 747, let alone a whole fleet of them, is a complex business. If you have the cash then an outright purchase is a possibility. If you don't, then the other option is to lease and most large aircraft in operation today are leased from banks.

If you are leasing the aircraft then there are contracts that have to be negotiated, covering everything from how much you are actually going to pay and when you are going to pay it, through to the aircraft's configuration (aircraft, like cars, have optional extras) and the maintenance schedule. It will also cover return conditions, which can sometimes require any modifications done by the operator to be removed, at the operator's expense.

There are a number of optional extras that the customer can choose. For the -400 the major options included:

- *A choice of engines:* P&W 4000 series, GE CF6-80, RR RB211-524.
- *All engine options*: Autostart system.
- *Fuel:* Horizontal stabiliser fuel tank.
- *Electrics:* TR for APU start.
- *Hydraulics:* Demand Pumps – air driven for systems 2 and 3; Electrical Auxiliary pump – system 1.
- *Flight controls:* Rudder trim centring switch.
- *Fire protection:* Automatic APU shutdown and extinguisher discharge.

ABOVE Etihad Airbus A340-600, A6-EHL.
(Chris Wood)

RIGHT Singapore Airlines Airbus A380, 9V-SKA.
(Chris Wood)

Airbus A340, which first flew in October 1991 and entered revenue service in February 1993. In April 1996 Airbus launched a stretched version, the A340-600, but in the early part of the 21st century the orders dwindled and the programme was terminated in November 2011 after a total of 377 had been built.

None of these aircraft came close to challenging the 747's supremacy in the long-haul market. However, the 747's dominance is now being threatened on two fronts – by Boeing's own twin-engine 777, particularly the stretched -300 and -300ER versions, but also by what is now the world's largest commercial aircraft, the Airbus A380.

It will be interesting to see if the 747-8 can regain the 747's pole position, or whether the economics of twin engines or bigger aircraft win the day.

RIGHT Egyptair Boeing 777-300ER, SU-GDN.
(Chris Wood)

Chapter Two

Anatomy of the Boeing 747

The Boeing 747 is a big aircraft in more ways than just its imposing appearance. Each 747 is made up of a staggering 6 million components, half of which are fasteners. It has 171 miles of wiring, 5 miles of tubing and uses 147,000lb of high-strength aluminium in its construction.

OPPOSITE Body gear oleo cylinder, wheels and steering actuator on a Boeing 747-400. *(Chris Wood)*

Construction

The 747 was so large that Boeing did not have anywhere big enough to build it, so they had to construct a new factory for the purpose. A site was chosen at Paine Field, Everett, some 30 miles north of Seattle and the resulting 200-million-cubic-foot building was so big that it features in the *Guinness Book of World Records* as the world's largest building by volume. It is also reported that it has its own microclimate, with clouds forming inside it at times! Such was the pace of development of

ABOVE The Boeing 747-100 production line at Everett. *(Boeing)*

RIGHT 747-400 production under way at Everett. The colours on the rudders indicate the airline customer. In the foreground Qantas, followed by British Airways and Korean Airlines. *(Boeing)*

BELOW The 75th 747 for Japan Airlines (JAL) was this 747-400, seen here in its final stage of assembly. *(Boeing)*

the 747 that the factory was still not finished when the first aircraft started rolling off the production line. Boeing 747s are still built at Everett, although it is more accurate to call it an assembly line rather than a factory as more than 50% of the parts are built elsewhere and shipped into Everett for assembly.

Each 747 is made up of a staggering 6 million components, half of which are fasteners. It has 274km (171 miles) of wiring, 8km (5 miles) of tubing and uses 66,150kg (147,000lb) of high-strength aluminium in its construction.

Most of the aircraft structure is built of aluminium alloy, a mix of 2000 series alloys (alloyed with zinc) and 7000 series alloys (alloyed with copper). Items requiring more strength use high-strength steel or titanium, and items that are not critical are made of a fibreglass honeycomb.

In 1992 it was estimated that the 747 programme provided work for 80,000 people, only 32,800 of whom were Boeing employees. The list of subcontractors for the 747 reads like a 'who's who' of the world's aviation companies. As an example, parts for the airframe assemblies come from such diverse places as Australia, China, Israel, Japan, Korea and the Netherlands.

Fuselage

The fuselage is constructed of circumferential frames and longitudinal stiffeners, giving it a semi-monocoque structure. The skin has fail-safe straps that act as tear stoppers. It has two floors in the forward part, and one in the rear, made up of beams and panels. The entire fuselage is pressurised, with the exception of the wheel wells and the centre wing box.

The fuselage is divided into sections, the front part to just behind the first main-deck

ABOVE A completed 747-400 leaves the construction hangar on its way to the paint shop at Everett. *(Boeing)*

LEFT The 747-400 fuselage is modelled here by British Airways' G-BNLZ taking off from Heathrow on 11 February 2010. *(Chris Wood)*

doors is known as Section 41. Section 42 goes as far as the rear of the upper deck. Section 44 is the centre section of the fuselage, which contains the centre wing box and the main landing gear wheel wells. Section 46 is the fuselage aft of the wing to the rear pressure bulkhead and Section 48 is the rear part of the fuselage, which contains the APU and the horizontal stabiliser centre section.

Most of the fuselage is constructed from aluminium alloys but the nose radome is fibreglass. All the landing gear is made of high-strength steel, the main gear being attached to the fuselage via a titanium beam. Titanium is also used for the APU firewall.

Wings

The wing area of the -400 is 524.9sq m (5,600sq ft) and the wing structure weighs 43,090kg (95,000lb). This is approximately 2,270kg (5,000lb) less than the wing of the Classic 747s, despite the wing being extended by 1.8m (6ft) either side and the addition of winglets. This is due to the use of improved aluminium alloys, which are stronger but lighter.

The wing consists of a left, centre and right wing box. The left and right wing boxes consist of a front, centre and rear spar, ribs, z-stringers and top and bottom skins. Additionally there are ailerons, spoilers, leading edge flaps, trailing

RIGHT A section through the fuselage of a 747 on display at the London Science Museum showing the positioning of the cargo hold, main passenger cabin and upper deck. *(Jonathan Falconer)*

ABOVE The upper section of the front pressure bulkhead inside a 747-400. *(Jonathan Falconer)*

FAR RIGHT The front pressure bulkhead inside the cargo hold of a 747-200F. *(Bas Tolsma/ TheDutchAviation.com)*

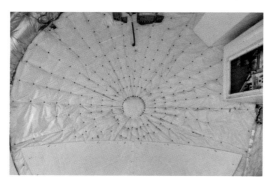

RIGHT Floor beams of the upper deck as seen from the main passenger deck below. *(Jonathan Falconer)*

FAR RIGHT Construction details of the fuselage wall. *(Jonathan Falconer)*

RIGHT 747-400 wing general view. This is Virgin's G-VROY high over the Atlantic.

BELOW Port wing structure, from the winglet – outboard aileron, outboard trailing edge flaps (with canoe fairings), inboard aileron, inboard trailing edge flaps. The spoiler panels are in front of the outboard flaps. *(Luis Santios/Shutterstock.com)*

LEFT Wingtip configuration of the 747 Classic showing outboard aileron, outer wing leading edge flap and static electricity discharge wick. *(Jonathan Falconer)*

FAR LEFT Lift spoilers in action on the upper wing surfaces of Virgin's G-VXLG landing at Las Vegas.

LEFT Leading edge and trailing edge flaps deployed on BA's G-CIVK on finals to Miami on 28 July 2009. *(Chris Wood)*

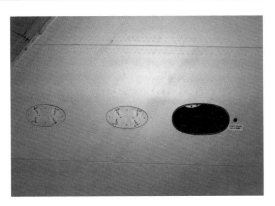

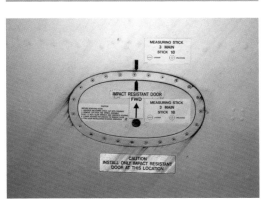

FAR LEFT Fuel tank access panels. *(Chris Wood)*

LEFT Fuel tank access panel and fuel measuring stick access. *(Chris Wood)*

Cutaway of the Boeing 747-400. *(Mike Badrocke)*

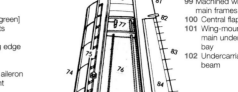

1 Radome
2 Weather radar scanner
3 Front pressure bulkhead
4 Scanner tracking mechanism
5 Wardrobe
6 First-class cabin, 30 or 34 seats at 62in [1.57m] pitch
7 Nose undercarriage wheel bay
8 Nose wheel doors
9 Twin nose wheels
10 Hydraulic steering jacks
11 Nose undercarriage pivot mounting
12 Under-floor avionics equipment racks
13 Cabin window panels
14 First-class bar unit
15 Flight deck floor level
16 Rudder pedals
17 Control column
18 Instrument panel, five CRT EFIS displays
19 Instrument panel shroud
20 Windscreen panels
21 Overhead systems switch panel
22 First Officer's seat
23 Captain's seat [two-crew cockpit]
24 Observer's folding seats [2]
25 Starboard side toilet compartments [2]
26 Cockpit bulkhead
27 Crew rest bunks [2]
28 Upper deck window panel
29 Conditioned air distribution ducting
30 Forward main deck galley unit
31 Plug-type forward cabin door, No. 1 port and starboard
32 Business-class passenger seating, 24 seats typical at 36in [91cm] pitch
33 Fuselage lower lobe skin panelling
34 Baggage/cargo pallet containers
35 Forward under-floor cargo hold, capacity 2,768cu ft [78.4cu m]
36 Forward fuselage frame and stringer construction
37 Upper deck doorway, port and starboard
38 Cabin roof frames
39 Anti-collision beacon light

40 No. 1 UHF communications antenna
41 Upper deck passenger cabin, 52 business-class or 69 economy-class seats
42 Lower deck sidewall toilet compartment
43 No. 2 passenger door, port and starboard
44 Air conditioning system heat exchanger intake ducting
45 Ventral ram air intakes
46 Faired wing root leading edge fillet
47 Ventral air conditioning packs, port and starboard
48 Wing spar bulkhead
49 Economy-class seating
50 Staircase to upper deck level
51 Fresh water tanks
52 Wing centre 4-section fuel tankage, capacity 16,990 US gal [64,315 lit]
53 Centre-section stringer construction
54 Floor beam structure
55 Front spar/fuselage main frame
56 Upper deck lobby area
57 Curtained bulkhead
58 Galley units
59 Starboard wing inboard main fuel tank, capacity 12,546 US gal [47,492 lit]
60 Fuel pumps
61 Engine bleed air supply ducting
62 Kruger flap operating mechanism
63 Inboard Kruger flap segments
64 Starboard inner Pratt & Whitney PW4256 engine nacelle
65 Inboard nacelle pylon
66 Leading-edge Kruger flap segments
67 Pressure refuelling connections, port and starboard
68 Kruger flap drive shaft

69 Kruger flap rotary actuators
70 Starboard wing outer main fuel tank, capacity 4,482 US gal [16,966 lit]
71 Starboard outer engine nacelle
72 Outer nacelle pylon
73 Starboard wing reserve tank provision, capacity 534 US gal [2,021 lit]
74 Outboard Krueger flap
75 Kruger flap drive mechanism
76 Outer wing panel dry bay
77 Vent surge tank
78 Wing-tip extension
79 Starboard navigation [green] and strobe [white] lights
80 Starboard winglet
81 Fixed portion of trailing edge
82 Fuel vent
83 Static dischargers
84 Outboard, low-speed, aileron
85 Outboard four-segment spoilers
86 Outboard triple-slotted Fowler-type flap, extended
87 Flap screw jacks and segment linkages

88 Flap drive shaft
89 Inboard, high speed, aileron
90 Inboard triple-slotted flap, extended

91 Inboard two-segment spoilers/ lift dumpers
92 Inboard triple-slotted flap, extended
93 Auxiliary trailing edge wing spar
94 Cabin air distribution ducting
95 Extended upper deck rear bulkhead
96 Upper deck floor beams
97 Air system cross-feed ducting
98 Conditioned air risers
99 Machined wing spar attaching main frames
100 Central flap drive motors
101 Wing-mounted outboard main undercarriage wheel bay
102 Undercarriage mounting beam

103 Central keel section
104 Pressure floor above wheel bay
105 Centre fuselage frame and stringer construction
106 Dual navigation antennae
107 Cabin wall trim panelling
108 Seat mounting rails
109 Main cabin floor panelling
110 Fuselage-mounted inboard, main undercarriage wheel bay
111 Hydraulic retraction jack
112 Cabin window panel
113 Overhead conditioned air distribution ducting
114 Economy-class seating, 302 to 410 seats at 34in [86cm] pitch
115 Overhead stowage bins
116 Sidewall toilet compartments, port and starboard
117 Central cabin galley
118 No. 4 passenger door, port and starboard
119 Rear cabin passenger seating
120 Rear cabin galley
121 Rear cabin air supply ducting
122 Fuselage sidewall stowage bins

123 Control cable runs
124 Central overhead stowage bins
125 Cabin roof panels
126 Ten-abreast economy-class seating
127 Rear fuselage frame and stringer structure
128 Rear cabin seating
129 Access ladder to upper deck crew rest area
130 Overhead cabin crew rest area, six bunks and four seats typical
131 Rear pressure bulkhead
132 Fin root fillet
133 Starboard trimming tailplane
134 Static dischargers
135 Starboard elevator
136 Fin leading edge structure
137 Two-spar fin box structure
138 Fin-tip fairing
139 VOR localiser antenna
140 Static dischargers
141 Upper rudder segment
142 Lower rudder segment
143 Rudder hydraulic actuators
144 Tail cone frame structure

145 Pratt & Whitney Canada PW901A auxiliary power unit [APU]
146 Tail navigation and strobe lights [white]
147 APU exhaust
148 Port elevator inboard segment
149 Outboard elevator segment
150 Static dischargers
151 Port trimming tailplane structure
152 Elevator hydraulic actuators
153 Long range tailplane integral fuel tank, capacity 3,300 US gal [12,492-lit]

154 Tailplane sealing plate
155 Aft fuselage framing
156 Fin root attachment joint
157 Tailplane centre section
158 Tailplane trim screw jack
159 APU high pressure air supply duct

160 Lower deck rear cabin toilet compartments
161 No. 5 passenger door, port and starboard
162 Rear fuselage window panel
163 Under-floor bulk cargo hold, capacity 1,000cu ft [28.3cu m]
164 Rear main cargo/baggage hold, capacity 2,422cu ft [68.6cu m]
165 Baggage cargo pallet
166 Fuselage lower lobe frame and stringer structure
167 Wing root trailing edge fillet composite structure
168 Fuselage-mounted main undercarriage pivot fixing
169 Trailing edge auxiliary spar
170 Main wheel leg breaker strut
171 Wing-mounted main undercarriage pivot fixing
172 Hydraulic retraction jack
173 Four-wheel inboard main undercarriage bogie
174 Flap drive shaft
175 Flap guide rails
176 Inboard spoiler panels/lift dumpers
177 Port inboard triple-slotted flap
178 Flap track fairings

179 Flap extended position
180 Aileron hydraulic actuator
181 Inboard, high-speed, aileron
182 Outboard triple-slotted flap
183 Outboard flap tracks
184 Outboard spoiler panels
185 Flap track fairings
186 Flap extended position
187 Outboard, low-speed, aileron

188 Aileron hydraulic actuators
189 Static dischargers
190 Fuel vent
191 Fixed portion of trailing edge
192 Port winglet
193 Winglet composite structure
194 Port navigation [red] and strobe [white] lights
195 Outboard leading edge Krueger flap segments
196 Krueger flap drive mechanism
197 Outer wing panel rib structure
198 Wing bottom skin access manholes
199 Rear spar
200 Outboard engine mounting rib
201 Port outer nacelle pylon
202 Thrust reverser cowling door

203 Reverser cascades
204 Outboard engine nacelle
205 Rolls-Royce RB211-524G alternative engine installation
206 Full length nacelle cowling
207 Internal exhaust stream mixer duct
208 Central leading edge Krueger flap segments
209 Krueger flap drive mechanism
210 Leading-edge rib structure
211 Outer wing panel three-spar torsion box structure
212 Wing ribs
213 Rear spar
214 Front spar
215 Wing stringers
216 Wing skin panelling
217 Wing-mounted main undercarriage leg strut
218 Pylon attachment strut
219 Four-wheel outer main undercarriage bogie
220 Nacelle pylon structure
221 Engine bleed air pre-cooler

222 Core engine, hot stream, exhaust duct
223 Fan air, cold stream, exhaust duct
224 Ventral engine accessory equipment pack
225 Pratt & Whitney PW4256 turbofan engine
226 Engine intake with acoustic lining
227 Detachable cowling panels
228 Bleed air de-iced intake lip
229 Inboard Krueger flap segments
230 Krueger flap motor and drive shaft
231 Machined spar booms
232 Inboard wing ribs
233 Bolted wing robot attachment joint strap
234 Front spar
235 Engine bleed air ducting
236 Leading edge nose ribs
237 Twin landing lamps
238 General Electric CF6-80C2 alternative engine installation

Mike Badrocke

51

edge flaps and winglets. Access panels in the undersides allow entry to the wing boxes, fuel tanks and also access to the fuel pumps.

The wing spars and skin are constructed from aluminium alloys, the upper and lower leading edge and trailing edge surfaces are made of a fibreglass honeycomb and the winglet from graphite composite with an aluminium leading edge.

The leading edge flaps, trailing edge flaps, aileron, spoilers and wing to body fairings are all fibreglass honeycomb.

ABOVE Standard 747 fin on Lufthansa's D-ABVF at Miami on 28 July 2009. *(Chris Wood)*

RIGHT SP fin on Air China's B-2454 at Gatwick on 21 May 1992. *(Chris Wood)*

Vertical stabiliser

The vertical stabiliser consists of front and rear spars, ribs, stringers and skin, which are all made from aluminium alloy. The rudders are a fibreglass honeycomb.

Horizontal stabiliser

The horizontal stabiliser is made up of left and right wing boxes. There are front and rear spars, with ribs, stringers and skin, all made from aluminium alloy. The elevators are attached to the rear spar and are a fibreglass honeycomb.

Both the horizontal and vertical stabilisers have access panels and internal crawl ways for inspection and maintenance.

RIGHT Horizontal stabiliser. *(Jonathan Falconer)*

Electronic equipment centres

The electronic equipment is fitted in various locations around the aircraft. Most of it is in the Main Equipment Centre (MEC), which is

RIGHT Nose wheel and ladder access to the MEC. *(Chris Wood)*

FAR RIGHT MEC forward access hatch. *(Chris Wood)*

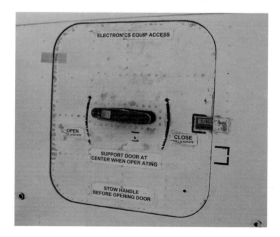

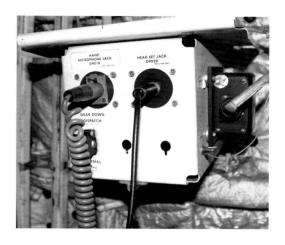

under the cabin floor, behind the nose wheel. It is normally accessed from a hatch in the underside of the fuselage. The majority of the black boxes in the MEC are mounted in two racks, running across the aircraft.

The Centre Equipment Centre is just forward of the wing and is accessible through a hatch on the underside of the fuselage. The Aft Equipment Centre is on the left-hand side in the rear of the main cabin. All the equipment is rack-mounted on shelves and is in black boxes. These are easily removed and replaced. (See Chapter 4 – Systems for further details.)

Flight controls

The aircraft has conventional flying controls in all three axes – pitch, roll and yaw. The primary controls for pitch are four elevators, with secondary control provided by a moveable and trimmable horizontal stabiliser. Primary roll control is given by four ailerons with spoilers for secondary control. Yaw control is provided by

ABOVE LEFT There are headset sockets around the aircraft for the engineers to communicate with each other at various locations, including the flight deck, whilst working on the aircraft. This one is in the MEC. *(Chris Wood)*

ABOVE Empty racking inside the forward MEC. *(Chris Wood)*

LEFT Avionics equipment in the MEC. *(Chris Wood)*

FAR LEFT Both the upper and lower rudders can be seen in this picture. *(Ian Black)*

LEFT Inboard elevator and moveable horizontal stabiliser. *(Ian Black)*

ABOVE High-lift devices are fitted to both the leading and trailing edges of the wing. All are visible in this photograph of Virgin's G-VXLG seconds from touchdown on RW 25L at Las Vegas.

two rudders. Each pilot has a control column for pitch control, which is fitted with a control wheel at the top for roll control. Each pilot also has a pair of adjustable rudder pedals for yaw control. All the controls are interconnected so that the aircraft can be flown by either pilot. Flying control surface position information can be displayed on the status page of the secondary EICAS (Engine-Indicating and Crew-Alerting System), and on the CMC (Central Maintenance Computer) pages. (For more information on EICAS see Chapter 4.)

High-lift devices consisting of flaps on both the leading and trailing edges of the wings provide additional lift, allow slow-speed flight for take-off, approach and landing. When fully extended the flaps increase the wing area by around 21% and the lift by around 90%. Flap position is displayed on the upper EICAS.

There is also a stall warning system.

All four hydraulic systems power the primary flying controls, with control split between the

systems such that the aircraft can still be flown using any one system. However, due to the size of the surfaces and the force required to move them there is no manual reversion, so at least one system is required. The leading edge flaps are normally pneumatically powered and the trailing edge flaps are normally hydraulically powered. They both have a secondary electrical power source in case of failure of the primary power source.

Pitch control

The four elevators are attached to the rear spar of the horizontal stabiliser. Inputs from the pilot's control columns are fed via a system of linkages and cables, through an elevator feel system, to hydraulic actuators attached to the inboard elevators. The outboard elevators are slaved to their respective inboard elevators via a slave linkage.

All four hydraulic systems power the elevators; systems 1 and 2 power the left and

RIGHT Control cable runs on the upper cabin ceiling of a 747-400. (Jonathan Falconer)

FAR RIGHT Exposed control cable runs in the cargo bay of a 747-200F. (Bas Tolsma/ TheDutchAviation.com)

systems 3 and 4 the right. The inboard elevators are powered by two systems through dual tandem power packages, the outboard by single systems through single power packages. The left outboard elevator is powered only by system 1 and the right outboard only by system 4.

The elevator feel system, which consists of a feel unit and a feel computer, provides artificial feel to the control columns to prevent over-pitching. The amount of feel depends on aircraft speed, with feedback force increasing as aircraft speed increases. The elevator feel system is powered by hydraulic systems 2 and 3 with a mechanical spring system as a back-up in case both hydraulic systems fail. In this event elevator feel is still active but is no longer a function of airspeed.

Autopilot inputs to the elevators are made through three modules mounted behind the stabiliser. Output from the modules is mechanically transmitted to the inboard elevator actuators by torque tube.

Pitch trim is provided through movement of the horizontal stabiliser through a range of 15°. Two electrical switches on each pilot's control column provide power to a pair of stabiliser trim control modules which control two independent hydraulic drive motors, one powered from hydraulic system 2 and one from system 3. One switch provides an arm signal, the other a control signal and both signals must be sensed before the stabiliser can move. This is to prevent unscheduled stabiliser movement. If the normal switches fail another pair of switches, the alternate stabiliser trim switches, mounted on the centre console can be used. If one of the hydraulic systems fails, stabiliser trim is still available but at half the normal speed.

Two guarded hydraulic cut-out switches are fitted on the centre console next to the alternate stabiliser trim switches, one for each hydraulic system. They have three positions; ON, AUTO and CUT OUT. They are normally in the AUTO position, which allows automatic cutting off of the related hydraulic supply if unscheduled movement of the stabiliser is sensed. ON overrides the automatic cut out whilst CUT OUT shuts off the hydraulic supply to the related stabiliser trim control module.

Stabiliser position is shown on trim indicators mounted on both sides of the control stand.

Roll control

The ailerons are divided into two pairs, inboard and outboard. The inboard pair are always active, the outboard pair are only active at low speeds; at higher speeds they are electrically locked out. Outboard aileron lockout occurs when the flaps are fully retracted and the indicated airspeed is greater than 238kt or M0.53. They are unlocked when the flaps are out of UP or the indicated airspeed is less than 232kt, or M0.51.

Each aileron has two hydraulic power sources and is driven by a dual tandem power control package. The left outboard aileron is powered by hydraulic systems 1 and 2, the left inboard by 1 and 3, the right inboard by 2 and 4 and the right outboard by 3 and 4.

Inputs from the pilot's control wheels are fed by cable to a trim, centre and feel mechanism in the left-hand wheel well. This mechanism has springs to provide artificial feel, which increase with control wheel movement. They also centre the control wheels and it is also where aileron trim signals are fed into the system. The output from this mechanism goes to two central lateral control computers (CLCP) that drive aileron programmers that are connected to the power control packages. Autopilot inputs are fed to the CLCPs.

Aileron trim is achieved through a pair of switches on the centre console. There is an arm switch and a control switch and both switches must be operated together for trimming to occur. Aileron trim position is indicated by the position of the control wheel, and maximum aileron trim is equivalent to 47° of control wheel movement (so it should be used with caution). Use of the aileron trim is not allowed when the autopilot is engaged.

Spoiler operation is either through the pilot's control wheels, or the speed brake lever which is mounted on the left-hand side of the centre console. There are a total of 12 spoiler panels on the trailing edge of the upper surface of the wings, six on each side, split into two groups either side. They are operated by hydraulic power control units, powered by hydraulic systems 2, 3 and 4, divided up symmetrically between the systems. System 2 powers panels 2, 3, 10 and 11, system 3 panels 1, 4, 9 and 10 and system 4 panels 5, 6, 7 and 8.

The outer five panels on each wing act as differential roll spoilers controlled by the pilot's control wheels. If the control wheel is moved by more than two units the spoiler panels will start to rise on their respective side. In the flight roll mode panels 1 to 4 and 9 to 12 move a maximum of 45°, whilst 5 and 8 can move a maximum of 20°.

When used as speed brakes in the air, panels 1, 2, 11 and 12 don't move, panels 3, 4, 9 and 10 move a maximum of 45°, and 5 to 8 a maximum of 20°. When used as speed brakes on the ground all the panels can move up to 45°. The amount of actual movement is determined by the position of the speed brake lever.

In the flight roll mode hydraulic output from the CLCPs feeds two spoiler differential mechanisms in the wing wheel wells. These are connected by cables to the power control units on the spoiler panels. The left spoiler differential mechanism powers panels 3, 4, 5 and their symmetrical pairs 8, 9 and 10, whilst the right spoiler differential mechanism powers panels 1, 2, 11 and 12.

The speed brake lever has four positions: DN, ARM, FLIGHT DETENT and UP. The DN (down) position moves all the panels down. However, if the main wheels are on the ground, thrust levers 2 and 3 are near the closed position and reverse thrust levers 2 or 4 are raised to the reverse idle detent, the speed brake lever is raised out of DN and moved to the UP position and ground spoilers deploy. This provides automatic speed brake deployment for the rejected take-off situation. The ARM position is used to arm the speed brake for landing; in this position if thrust levers No 1 and 3 are near the closed position, when the main wheels touch the ground the ground spoilers deploy automatically. The FLIGHT DETENT position is as far as it is possible to move the lever in flight, the UP position is only available on the ground. The lever operates all the panels symmetrically via a cable to the speed brake sequence mechanism. This feeds to the spoiler differential mechanism and also to a ground spoiler control valve in the right wing wheel well.

Yaw control

There are two rudders, an upper one and a lower one, controlled by two interconnected pairs of rudder pedals in the flight deck, one pair for each pilot. The two rudders are similar in design, have similar capability and operate together. The system also includes a feel system, a trim system, a pair of ratio changers (one for each rudder), and a pair of yaw dampers (one for each rudder).

Inputs from the rudder pedals are fed by cable to a single feel, centring and trim mechanism. From here they are fed by control rods via the ratio changers to a pair of power control modules. The power control modules also receive inputs from the yaw dampers. The output from the power control modules is fed to five power control actuators. The upper rudder is powered by Nos 1 and 3 hydraulic systems, and is driven though a dual hydraulic triple actuator power unit. The lower rudder is

BELOW Trailing edge flaps deployed at the flap 30 position. *(Jonathan Falconer)*

powered by Nos 2 and 4 hydraulic systems, and is driven through a dual hydraulic dual actuator power unit.

The feel, centring and trim mechanism has springs to provide feel, to centre the pedals and to set the pedal neutral position by use of the rudder trim. Rudder trim is controlled electrically using a large round knob on the control stand from where signals are fed to an actuator on the feel centring and trim mechanism. Trim position is shown on an adjacent indicator; the maximum amount of trim that can be applied is 80% of rudder travel. The rudder trim can be set to zero with a rudder trim centring switch which is next to the rudder trim knob.

The rudder ratio changers reduce the amount of rudder travel for a given amount of pedal movement as airspeed increases. For this they receive airspeed inputs from the digital air data computers. Below 150kt the rudder has full authority and can travel ± 31.5°. At high speeds, above 450kt, movement is restricted to ± 4.5°.

The yaw dampers provide inputs to the power control modules to oppose any yawing due to Dutch Roll. They also provide turn coordination and help to reduce the lateral flexing caused by turbulence. The system uses signals from the Inertial Reference System (IRS), the Air Data Computers (ADC) and four accelerometers. The yaw dampers move the rudders, but not the rudder pedals.

High-lift devices

High-lift devices consist of flaps on both the leading and trailing edges of the wings. There are three modes of operation: primary mode, secondary electrical mode and alternate electrical mode. In the primary and secondary modes their position is selected by the flap lever, for the alternate mode there are two switches on the forward panel, one to arm the system and one to activate it. In all modes both the leading and trailing edge flaps operate together.

Primary mode sees the leading edge flaps operated pneumatically, using bleed air from the pneumatic manifold which is supplied though ducts in the wing leading edge. The trailing edge flaps are powered hydraulically, with system 1 powering the inboard flaps and system 4 the outboard. If pneumatics or hydraulics are not

available the secondary electrical system takes over and operates the affected surfaces.

Available flap positions are UP, 1, 5, 10, 20, 25 and 30. Flap 10 and 20 are the normal take-off flap settings; Flap 25 and 30 are the normal landing flap settings.

The flap position is shown on the upper EICAS display. It normally shows as a strip, in magenta when the flaps are travelling to the selected position, and in green once they get there. This display blanks 10 seconds after the flaps are retracted. In the event of a flap malfunction an expanded display appears automatically, showing the position of each individual surface.

The flaps are controlled by three Flap Control Units (FCUs) located in the MEC. The FCUs have three functions: primary control, secondary electrical control and indication. The control functions are split into: extension and retraction for both leading and trailing edge flaps and asymmetry protection, failure protection and load relief for the trailing edge only. The indication function displays position on the EICAS, activates EICAS messages for failures and sends signals to the CMC. It also provides signals to other systems that require knowledge of the flap position.

Normally the left FCU controls the primary mode and the right controls the secondary electrical mode. All three monitor for failures. In the event of an FCU failure control is automatically passed to another FCU and only one FCU is needed to perform all the functions. In the event of all the FCUs failing, control of the flaps is achieved through the alternate electrical mode. The main difference from the pilot's perspective between pneumatic and electrical

ABOVE Leading edge flaps deployed on BA's G-BNLS landing at Heathrow on 3 September 2009.
(Chris Wood)

RIGHT A view along the wing showing all three sets of leading edge flaps extended. Inboard are the Krueger flaps; the mid and outer sections are variable camber. *(Chris Wood)*

FAR RIGHT Variable camber mechanism. *(Chris Wood)*

RIGHT Detail of variable camber mechanism. *(Chris Wood)*

BELOW Leading edge flap drive unit showing the pneumatic motor on the right side and the electric motor on the left. *(Chris Wood)*

operation is that the flaps travel much more slowly when driven electrically.

The flap lever is mechanically connected to three rotary variable differential transformers (RVDTs) on the flight deck floor under the centre console, which sense movement and send an electrical signal to the three FCUs. The FCUs compare these signals with flap position signals and if a difference is sensed, they command movement to the desired flap setting.

Leading edge flaps

There are 28 leading edge flaps, 14 on each wing. On each side the inboard 3 are Krueger flaps, the remaining 11 are variable camber flaps that are split into 3 groups: the inboard section (3 panels), the mid-span section (5 panels) and the outboard section (6 panels). When stowed all of the leading edge flaps are flush with the underside of the wing. The variable camber flaps contain a mechanism that curves them as they are deployed, and both they and the Krueger flaps have a folding nose.

The leading edge flaps have two positions, UP or DOWN, and they are moved by eight drive units, four on each side. The mid-span sections have two drive units, the other sections have one each. Each drive unit has two motors, a pneumatic one and an electric one, and they drive rotary actuators that move the flaps.

In primary and secondary mode the FCU sends a signal to the drive units; in alternate mode the signal comes directly from the alternate switches.

Trailing edge flaps

The trailing edge flaps are in four sections, two on each wing, in two inboard and two outboard

ABOVE Trailing edge flaps showing the three sections. *(Chris Wood)*

ABOVE Trailing edge flaps in operation. *(Chris Wood)*

LEFT Trailing edge flaps and canoe fairings that cover the flap-drive mechanism. *(Chris Wood)*

FAR LEFT Flap-drive mechanism. *(Chris Wood)*

LEFT Inboard trailing edge flap power package inside the left-hand body gear wheel well. *(Chris Wood)*

symmetrical pairs. They are triple-slotted, which means that each flap section is actually made up of three parts, a fore flap, a mid-flap and an aft flap, connected by linkages. Each section has two flap tracks attached to the underside of the wing and the mid-flap. The flap tracks are covered by an aerodynamic fairing known as the canoe fairing.

Signals from the FCUs are supplied to the flap power package for the inboard flaps, mounted in the left-hand body gear wheel well. This is connected by a linkage to the flap power package for the outboard flaps in the right-hand body gear wheel well. The power packages have two motors, one hydraulic and the other electrical. The motors drive torque tubes, which are connected via transmissions to ball screws that rotate and move the flap ball screw nut along the ball screw, moving the flaps to the selected position.

Flap operation

When the flap lever is moved from UP to Flap 1, the inboard and mid-span groups of the leading edge flaps deploy. When Flap 5 is selected the outboard group of leading edge flaps deploy and the trailing edge flaps move to the Flap 5 position. Selection of Flap 10, 20, 25 and 30 results in only the trailing edge flaps moving. The reverse happens during flap retraction.

Flap limiting speeds are placarded on the flight deck forward panel. Minimum speeds for flying at the different flap settings are calculated by the FMCs and displayed on the PFD speed scale (they are a function of the aircraft's weight). If the limiting speed for Flap 25 or 30 is exceeded with that flap selected, the flap load relief system automatically retracts the flaps, up to Flap 20. This protects them from excessive air loads. When the airspeed is reduced the flaps return to the selected position. Flap load relief is only available in the primary mode.

In the alternate mode, when extension is selected all the flap groups start travelling. The trailing edge flaps only travel as far as the Flap 25 position. For retraction the inboard and mid-span leading edge flaps do not retract until the trailing edge flaps have fully retracted.

If the FCUs detect a flap asymmetry (ie one section moving one side but not the other) the FCU stops operation of that asymmetric group (inboard or outboard); the other group continues to the selected position. Asymmetry protection is not available in alternate mode.

When pneumatically operated it takes 9sec for the leading edge flaps to move through their full range. When electrically operated it takes 90sec, and around 4min for the trailing edge flaps to fully travel.

When reverse thrust is selected, the inboard and mid-span leading edge flaps are automatically retracted.

Stall warning

The aircraft has a stick shaker attached to each control column. There are two stall warning computer cards in the Modular Avionics Warning Electronic Assembly (MAWEA) in the MEC, which receive airspeed, angle of attack, pitch rate and flap position signals. They activate the stick shakers and provide information to other systems to generate warnings.

Landing gear

The 747 has a unique undercarriage arrangement with four main landing gears and a single nose gear. Each main landing gear has four wheels, each with its own brake unit making a total of 16, while the nose gear has two wheels that do not have brakes. The body gear and the nose gear are steerable on the ground, to aid turning when taxying or towing.

Gear retraction, extension, steering and wheel braking are all hydraulically powered, with the nose and body gear powered by the No 1 system and the wing gear by the No 4. In the event of hydraulic failure there is an alternate, electrically controlled, system for lowering the gear.

RIGHT Flap 20 set for take-off.
(Ian Black)

Main landing gear

Each main gear consists of a shock strut (an air/oil shock absorber) mounted on a trunnion attached to the aircraft structure. Attached to the bottom of the shock strut is a truck beam with a one-piece axle at either end; the wheels and brake units are fitted to the axles. A system of braces and struts hold the shock struts in place.

The gear is raised and lowered by hydraulic gear actuators. Downlock actuators keep the gear locked down and also unlock them for raising.

All the main gears tilt, aft wheels down, to allow them to fit in the wheel wells and also for air/ground sensing. The body gear is tilted 7° whilst the wing gear is tilted 52°, by hydraulic truck positioning actuators.

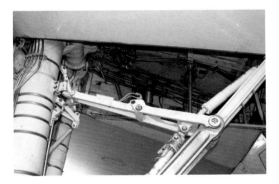

Nose landing gear

The nose landing gear consists of a shock strut mounted on a trunnion, which is attached to the aircraft structure. Attached to the bottom of the shock strut is a single dual-wheel axle. The two wheels are attached to the axle. The nose gear is maintained in either the up and locked, or down and locked, position by a drag strut assembly and a lock actuator.

Gear doors

All the landing gears have doors; each main gear has one hydraulically actuated wheel well door and one mechanically actuated shock strut door, whilst the nose gear has two of each. During the retraction or extension cycle the hydraulically actuated doors open to allow the gear to move

and then close; the mechanically operated doors are actually attached to the shock struts so stay open when the gear is down. The hydraulically operated doors can also be opened on the ground for maintenance access to the wheel wells.

Operation

The gear is normally raised and lowered using a single gear handle fitted in the pilot's front panel. It has three positions, UP, OFF and DN (down) and is easily identifiable by a small wheel at its end. On the ground there is a mechanical lock over the lever to prevent its inadvertent movement. This lock is automatically released when the aircraft gets airborne (or more accurately when the safe conditions for retraction are met, ie body gear centred and all main gear tilted).

When selected UP, hydraulic power is supplied to the system, via sequence valves to control the operation. First the wheel well doors open, then the gear downlocks are released by their hydraulic actuators, the gear is raised by its actuators and finally the wheel well doors close. The main gears are held in place by mechanical uplocks; the nose gear is held in place by its lock actuator. When the sequence has finished the gear handle can be moved to the OFF position to depressurise the system. When retracted the main gear is automatically braked to stop the wheels rotating; the nose

FAR LEFT Main gear doors. *(Jonathan Falconer)*

LEFT Inside the main gear bay. *(Jonathan Falconer)*

wheels are braked by rubbing strips in the nose wheel bay.

When the lever is moved to the DN position, the wheel well doors open, the uplocks are released and the gear falls under gravity to the down and locked position. The downlock actuators ensure the gears stay in the locked position. The wheel well doors are then closed and the main gears are positioned into their tilted positions by the truck positioning actuators.

Landing gear alternate extension

If the normal extension system is not available (ie, in the event of hydraulic system failure) there is an alternate electrical system, controlled by a pair of switches on the pilot's forward panel, one for the nose and body gear (normally powered by hydraulic system No 1) and one for the wing gear (normally powered by hydraulic system No 4). When the switches are pressed electric actuators are powered to unlock the main gear wheel well doors and uplocks, the nose gear doors and a lockshaft to unlock the nose gear. All the gears fall under gravity and are locked by the downlocks, which are pulled into position by springs.

The wheel well doors stay open when the alternate gear extension system is used and the aircraft can be landed with them open. It can actually be landed on any two main gears (plus the nose gear obviously), provided there is at least one down on each side, and the runway is long enough to cope with the reduced braking capability.

Landing gear indication

Landing gear position is shown on the upper EICAS screen. When it is down and locked the indication is a green box with DOWN in the middle; when the gear is in transit the box turns white and is cross-hatched; when the

gear is up it stays white with UP in the box. However, after 10sec the display blanks. If there is a malfunction an expanded display appears showing each individual gear and its status.

Nose gear steering

The nose gear is steerable, up to ± 70° using a pair of tillers, one per pilot, mounted on the pilot's sidewall. It can also be controlled through the rudder pedals, up to ± 7°, although this is overridden by tiller inputs. The tillers are connected by cables to steering metering valves which control the flow of hydraulic fluid to a pair of actuators mounted on the shock strut. Nose wheel steering is powered by the No 1 hydraulic system. For pushback the nose wheel steering can be bypassed with a steering bypass pin.

LEFT Nose wheel shock strut and steering cables. *(Chris Wood)*

LEFT Nose wheel steering actuators. *(Chris Wood)*

RIGHT Body gear steering actuator and wheels. (Chris Wood)

RIGHT Body gear steering actuator. (Chris Wood)

Body gear steering

The body gear steering is activated automatically by a switch on the nose gear, when the nose wheel steering angle exceeds 20° and the aircraft's speed is less than 15kt. An electronic control unit modulates the steering commands to the hydraulic actuators so body gear steering is proportional to nose wheel steering. The maximum angle, 13°, is achieved when the nose wheel is at its maximum angle of 70°. It is deactivated when the aircraft's speed increases above 20kt, or the nose wheel steering angle is less than 20°. The body gear turns in the opposite direction to the nose wheel, and like the nose wheel steering is powered by No 1 hydraulic system. Both nose wheel and body gear steering can only be activated when the aircraft is on the ground.

Air/ground sensing

Located in the MEC is a device known as the Proximity Switch Electronic Unit (PSEU). This receives signals from sensors at the landing gear and landing gear doors and provides data to the EIUs, the CMCS and the EICAS, as well as to air/ground relays, also in the MEC. These use tilt signals from the main gear trucks and an extension signal from the nose gear to sense whether the aircraft is on the ground or in the air. This is used to control various system functions that are configured with ground and air modes. All the sensors are dual to provide redundancy.

Wheels and brakes

All the wheels are two-piece forged aluminium assemblies. Each 22in-diameter main wheel has a heat shield and three thermal fuse plugs, which are designed to deflate the tyre rather than allow it to explode if it gets overheated. Each wheel also has an over-pressure relief valve. The nose wheels are normally 20in diameter, but 22in main wheels can be fitted to the nose gear (provided that a matching pair is fitted).

Each main wheel is equipped with its own hydraulically actuated carbon disc brake unit. The brake unit is fitted with wear indicator pins, bleed fittings and automatic piston adjusters. The normal brake system is powered by the No 4 hydraulic system, and there is an alternate brake system that is usually powered by system No 1, with system No 2 as a back-up. The changeover between systems is automatic.

The brakes are generally applied using brake pedals fitted to the top of both pilots' rudder pedals. They are connected by cables to brake-metering valve modules, containing normal and alternate metering valves. These are located in

RIGHT Wing gear truck beam. (Chris Wood)

FAR RIGHT Wheel unit. (Jonathan Falconer)

the wheel wells. The brakes can also be applied by the autobrake system.

There is a parking brake operated by a lever next to the centre console. The parking brake is set by pressing both brake pedals fully down and pulling back the parking brake lever until it latches. The brake pedals are mechanically latched in the down position and the parking brake valve is closed, trapping hydraulic fluid in the brake lines. The parking brake is released by pressing down on the brake pedals.

The aircraft has a Brake Temperature Monitoring System (BTMS), which has temperature probes fitted in each brake unit and displays brake temperatures on the lower EICAS gear page.

Tubeless tyres are fitted to all the wheels. There is an optional Tyre Pressure Indication System (TPIS) that displays tyre pressures on the lower EICAS gear page.

Anti-skid

The aircraft has an anti-skid system, which controls the hydraulic pressure applied to each brake to prevent skidding. It works in both the normal and alternate braking modes with each gear truck having two sets of anti-skid valves, one for each system. The normal braking system controls each wheel individually; the alternate system controls wheel pairs on each axle. If a skid is detected in a wheel, brake pressure is reduced by the associated anti-skid valve until the skidding stops.

The anti-skid system also provides touchdown, hydroplaning and locked wheel protection.

Brake torque limiters

Each main wheel has a brake torque sensor. If this detects excessive torque being applied to the wheel during braking it sends a signal to the wheel's anti-skid valve to release brake pressure to that wheel.

Autobrakes

The autobrake system applies the brakes to achieve a predetermined deceleration rate with the normal braking system. It is not available with the alternate braking system. It has several landing settings: 1, 2, 3, 4 and MAX, and one take-off setting: RTO (Rejected Take-Off). It is selected ON with a rotary switch fitted on the centre console or the pilots' forward panel.

LEFT **Wheel brake unit.** *(Chris Wood)*

In the landing case the autobrakes are activated when the aircraft is on the ground (air/ground sensing), the wheels have spun up and the thrust levers are closed. It modulates to achieve the selected deceleration rate, so if other braking services such as reverse thrust are applied, the brakes will back off. The autobrakes can be cancelled by pressing the brake pedals, by moving the switch to OFF, by shifting the speedbrake lever to the DOWN detent, or by advancing any thrust lever. If it is not cancelled it will bring the aircraft to a stop.

For take-off the autobrakes are set to RTO. This is activated on the ground above 85kt groundspeed if all the thrust levers are closed. It applies maximum brake pressure and if not cancelled will bring the aircraft to a stop. During a normal take-off once the aircraft's air/ground sensing system detects that it is airborne, the autobrake selector moves to the OFF position.

Brake accumulator

Fitted in the right-hand body gear wheel well is a brake accumulator, pre-charged to 750psi. This is pressurised by the No 4 hydraulic system and provides pressure for the parking brake when the normal and alternate brake systems are not pressurised. A gauge on the Captain's instrument panel shows accumulator pressure.

Windows and doors

Windshields

All the flight-deck windshields are electrically heated. The two forward windows have switches and are anti-iced and anti-fogged. The four side windows are permanently on for anti-fogging. They all have automatic temperature control.

There is a pair of two-speed windshield wipers, one per pilot, controlled by a pair of switches on the flight-deck overhead panel. The switches have three positions: OFF, LO (160 strokes/min) and HI (250 strokes/min).

There is a windshield washer consisting of a reservoir in the book stowage behind the Captain's seat, a pair of nozzles below the two forward windshields and a pair of switches by the wiper switches.

The aircraft can be fitted with a rain repellent system, which is designed to improve visibility during heavy rain. The repellent is housed in a can fitted in the flight deck and it is selected using a pair of switches above the wiper switches.

Doors

The aircraft is fitted with a selection of doors: 10 passenger doors on the main deck for embarking and disembarking passengers, 2 emergency escape doors on the upper deck, 3 cargo doors, 2 electrical and electronic equipment centre access doors, and a flight deck escape hatch. All twelve of the cabin doors are fitted with emergency escape slides. With the exception of the two upper-deck doors and the two main cargo doors, all the doors are plug-type doors, ie like bathroom plugs they are bigger than the holes they plug.

The position of all the doors is shown on the

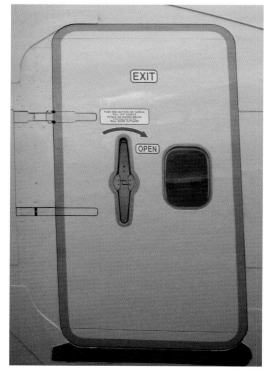

RIGHT **Upper-deck door, inside view.** *(Chris Wood)*

FAR RIGHT **Main-deck door, external view.** *(Ian Black)*

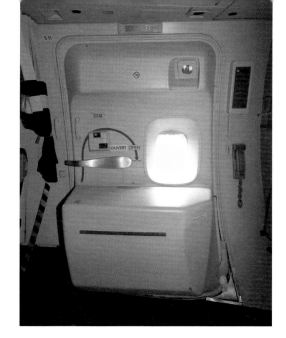

Doors synoptic of the lower EICAS. The position of the door arming lever can also be shown.

The ten main-deck doors are numbered from 1 to 5 and L and R (left and right), so door L1 is the forward door on the left side and R5 is the rear door on the right. As stated above they are plug-type doors; however, these doors actually open outwards and benefit from a clever design that allows them to do that. When opening them the initial movement is inwards and upwards, followed by an outwards and forwards motion, until locked into the open position. They can be operated from inside or outside in wind speeds of up to 40kt, and can be left open in winds of up to 65kt.

Each door has a mode selector lever. This is normally selected to Manual on the ground and Automatic for departure. In Automatic, rotating the door handle through 180° starts the door opening and activates the power-assist system. This utilises pneumatic pressure from a nitrogen reservoir to assist with door opening in an emergency.

The nitrogen reservoirs are located above doors 1, 2 and 4 and on the floor by doors 3 and 5. As the door opens the emergency escape slide is activated and inflates. The slides at doors 1, 2, 4 and 5 extend away from the aircraft and are designed such that in the event of a ditching they can be used as rafts (after they have been detached from the aircraft). The slides at the 3 doors are off-wing slides; they go down the wing along the side of the fuselage to the trailing edge and would interfere with the slides from the 4 doors in a ditching, so are

not used in that event (ie, the 3 doors mode selector levers would be selected to Manual.)

If the selector lever is in Automatic and the door is opened from the outside, it is mechanically switched back to Manual to prevent slide inflation.

The two upper-deck doors are for emergency evacuation only and they are fitted with slides, which are not suitable for use as rafts. The doors are hinged at the top, open outwards and can be opened from inside or outside. They are electrically powered for

normal operation and pneumatically powered for emergency operation, through a pair of motors (one electric, the other pneumatic) on the top of each door. These drive the door to the desired position.

For normal operation there are electrical switches both inside and outside. For emergency operation there is an operating handle inside and an electrical switch outside. Each door has a pair of Arm/Disarm levers, one inside and one outside, and an air bottle, charged to 3,000psi fitted with a cartridge. When the door is opened in an emergency the cartridge is electrically fired, discharging the bottle to assist with opening the door. The cartridge firing is normally powered by the aircraft's electrical system, but has its own battery for situations where that is not available.

The upper-deck doors have electrically powered flight lock actuators. These inhibit operation of the door handles when the aircraft is airborne. There is a blue DOOR GND MODE light above the door that illuminates when the actuator is in the unlocked position.

There are three cargo doors, all fitted on the right-hand side of the aircraft. The forward and rear cargo doors open outwards, and they are normally operated electrically using switches fitted outside in front of the door, or inside in front of the door. In the event of a failure of the normal system they can be operated manually. For this purpose there are three socket access ports, two in the centreline of the door and one above it. However, this is a slow and tedious job.

These doors are locked manually using the master lock latch handle fitted towards the front of the door. This handle also operates a pair of negative pressure relief doors in each main cargo door, to provide pressure equalisation. The doors have six access panels, which can be removed to gain access if the door fails in the closed position, and there are eight viewing windows for visually checking the latches.

The bulk cargo door opens inwards and is operated manually.

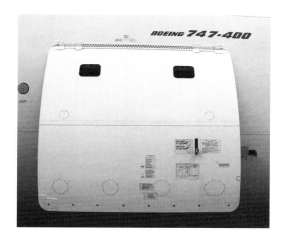

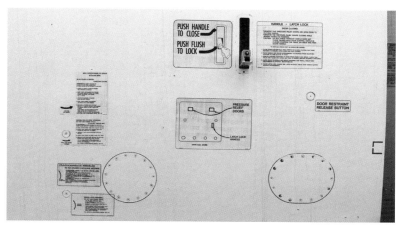

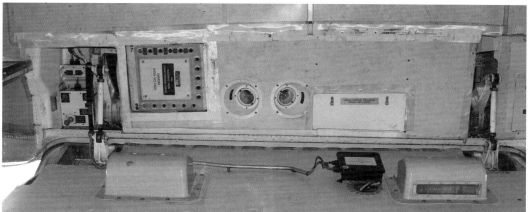

ABOVE LEFT Forward cargo door with the negative pressure relief doors open. *(Chris Wood)*

ABOVE Cargo door master lock latch handle. *(Chris Wood)*

LEFT Forward cargo door, view from inside cargo hold. *(Chris Wood)*

FAR LEFT Cargo door locks. *(Chris Wood)*

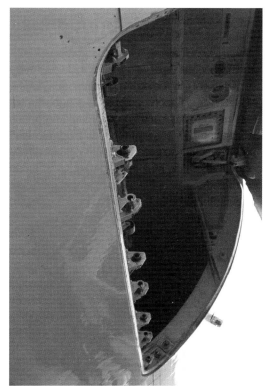

LEFT Interior view of rear cargo hold. *(Ian Black)*

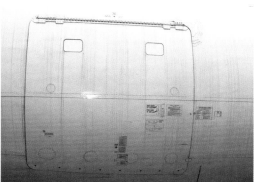

LEFT Rear cargo door with the negative pressure relief doors closed. *(Chris Wood)*

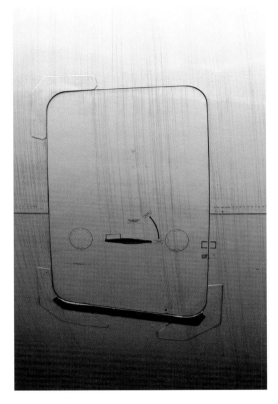

ABOVE Bulk cargo door. *(Chris Wood)*

LEFT 747-400F with upward-hinging nose for freight loading. *(Mikael Damkier/Shutterstock.com)*

RIGHT Empty forward cargo bay interior in a 747-400. *(Jonathan Falconer)*

FAR RIGHT Empty 747-200 Combi cargo hold interior. *(Bas Tolsma/ TheDutchAviator.com)*

RIGHT Cargo hold floor detail. *(Jonathan Falconer)*

FAR RIGHT Cargo hold bin locks on the floor. *(Chris Wood)*

The flight deck has a lockable security door separating it from the passenger cabin. There is also an overhead escape hatch that can be opened from the inside or the outside.

The -400F and -400ERF have an electrically operated nose cargo door that opens upwards. These models, the Combi and other 747s converted to freighter configuration have an electrically-operated, outwards-opening side cargo door. These are powered by the main deck cargo handling bus.

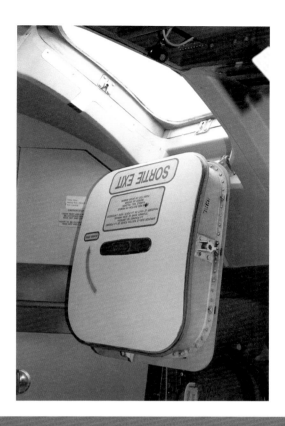

LEFT Flight-deck overhead hatch. *(Chris Wood)*

Lighting

External lights

The aircraft has various external lights to aid visual identification and to help the pilots see where they are going. There are pairs of navigation lights on each wingtip – red on the left wing and green on the right, and a pair of white lights at the extreme rear of the fuselage, mounted just below the APU exhaust outlet. There are also white strobe lights fitted in the

WINDSHIELD DESIGN

Aircraft windshields have precise requirements in use – resistance to extremes of temperature, fluctuations in cabin pressurisation, and resilience to high velocity bird impact. In addition, they must offer excellent visibility – both optically and for de-icing – and an aerodynamic design.

The Boeing 747 ushered in a new era in commercial airliner design and with it a new age in aircraft windshield design. All previous commercial airliners had used flat laminated glass windshields. Boeing adopted the curved windshield concept for the 747 at an early stage in the design based on two distinct advantages – better visibility and superior aerodynamics (including less drag and noise).

Triplex developed the 'Ten-Twenty' curved glass windscreen especially for the Boeing 747. Two pre-curved 12mm plies of Triplex 'Ten-Twenty' glass are laminated with plasticised polyvinyl butyral (PVB) and covered with a 3mm pre-curved facing ply of 'Ten-Twenty' glass. An electrically conductive coating, 'Hyviz', is applied to the inner surface

of the outer ply for de-icing and de-misting and is then laminated to the 12mm 'Ten-Twenty' ply.

The complete windscreen, weighing about 64kg, is clamped into the frontal area of the flight deck, protected from airframe structural deformation and the low temperatures achieved during high-altitude cruise.

BELOW Boeing 747-400 windshield. *(Swamysk/iStock)*

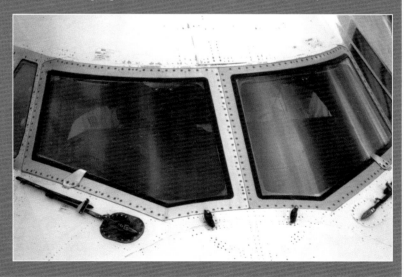

same positions and a pair of red anti-collision beacons – an upper one on top of the forward fuselage and a lower one underneath, forward of the landing gear bays.

In the wing leading edge close to the fuselage on each side is a pair of landing lights – an inboard pair and an outboard pair. These illuminate the area in front of the aircraft. When the landing gear lever is DOWN they are at maximum brightness, when the lever is UP or OFF they are dimmed.

Inboard of the landing lights, mounted in the fuselage, is a pair of wing lights. These illuminate the leading edges of the wings and the engine nacelles and can be used to check for icing.

Mounted on the nose wheel are two runway turn-off lights, which are offset at 65° to illuminate the area either side of the aircraft whilst taxying. There is also an option for a pair of taxi lights, which illuminate the area ahead of the aircraft. These lights are only powered when the aircraft is on the ground.

There are two logo lights, one on each horizontal stabiliser. These illuminate the vertical stabiliser.

Controls for all the external lights are on the pilot's overhead panel in the flight deck.

Cockpit lighting

All the cockpit panels are lit by integral lighting and they also have floodlights. Overhead dome lights provide illumination of the flight deck. They are controlled by switches on the overhead panel and the pilots' side panels.

Also on the overhead panel is an indicator lights switch. This has three positions: DIM, BRT, and TEST. DIM and BRT control the intensity of the annunciator lights (which are the lights in switches), TEST illuminates all of them at full brightness. They have integral bulbs, so it is always worth checking the bulb first, rather than assuming a system has failed.

Passenger cabins

There is a wide variety of possible cabin arrangements that can be used, depending on the requirements of the operator. Seats, galleys and lavatories are all removable, so the configuration can be changed relatively quickly. Every operator has its own seating configuration; in fact operators may have more than one for their fleet if their route structure dictates that requirement. They may also change over time as market conditions alter, and the aircraft has been designed with this factor in mind.

LEFT Passenger cabin windows are triple-glazed. *(Shutterstock.com)*

ABOVE Club World cabin on a BA 747-400. *(Gary Bembridge/ Wikimedia Commons)*

BELOW LEFT Economy class cabin on a United Airlines 747-422. *(Paulo Ordoveza/Wikimedia Commons)*

BELOW Business-class seating on a Singapore Airlines -400. *(Jordan Tan/Shutterstock.com)*

A typical business route will see a lot of First- and Business-class seats and relatively few Economy ones, so the total seating capacity could be around 350, whereas a leisure route will have few if any First- and Business-class seats and may be entirely Economy, with around 500 seats. However, the passenger aircraft is actually certified for a total of 660 passengers, 550 on the main deck and 110 on the upper deck.

(The Israeli airline El Al holds the record for the most people on board an aircraft, having squeezed 1,087 passengers into a 747-200 when evacuating people from Ethiopia in May 1991.)

The aircraft is fitted with an advanced cabin entertainment system (ACESS), which consists of the cabin interphone system, the passenger address system, the passenger entertainment system, the passenger service system and the cabin lighting system. The ACESS is an integrated system and it is controlled by a central management unit. It is designed to cope with configuration changes.

The aircraft has a cart lift system for moving galley carts to and from the upper deck from the main-deck galley between the L2 and R2 doors. This is electrically powered, using a pair of electrical motors. If one fails the other will still drive the lift; if both fail the lift can be operated

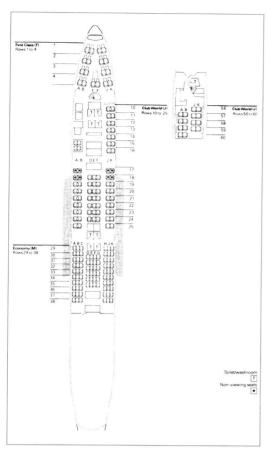

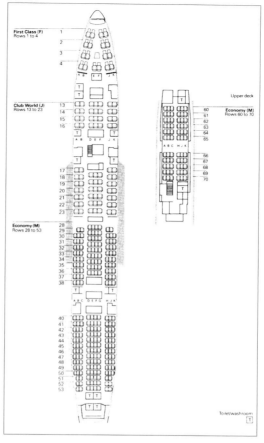

manually. There are two control panels, one on the upper deck and one on the main deck.

Stowage bins are fitted on the cabin sidewalls on both the main and upper decks and along the centreline of the cabin roof on the main deck. There are also side storage bins along the sidewalls of the upper deck.

There are 39 possible lavatory locations around the aircraft, and they can be fitted on both the main and upper decks. Each lavatory has a vacuum flush toilet, a washbasin and

lighting, with additional lighting being provided when the door is locked.

Crew rest areas are provided for both pilots and cabin crew. The pilots' rest area is inside the flight deck and contains a pair of bunks, plus an optional seat. Drop down oxygen masks are provided, and there is a light. The cabin crew rest area is above the rear of the cabin and is accessed via a locked door. It contains eight bunks and has oxygen masks, lighting and emergency equipment. There is also an emergency escape hatch.

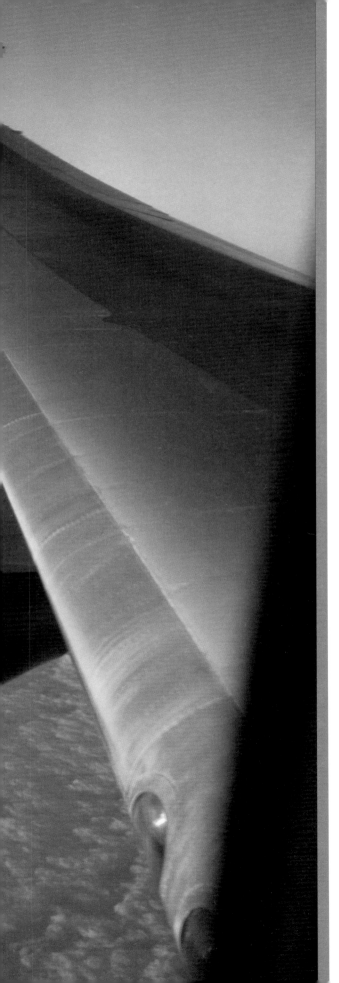

Chapter Three

Engines

The 747-400 comes with a choice of three different engine fits – the Pratt and Whitney PW 4000 series, General Electric CF6-80 series and the Rolls-Royce RB211-524 series. From the pilot's point of view the different engines present few differences in the way the aircraft is operated, neither is it a problem to have a 747 fleet with mixed powerplants.

OPPOSITE Virgin's 747-400s are fitted with General Electric CF6-80C2 engines. *(Ian Black)*

The 747 has four engines fitted in pods mounted under the wings. The engines are numbered from 1 to 4, with number 1 being the outboard left-hand engine, number 2 inboard left, number 3 inboard right and number 4 outboard right.

There is a choice of three different engine fits for the 747-400 – Pratt and Whitney's PW 4000 series, General Electric's CF6-80 series and Rolls-Royce's RB211-524 series. The versions normally fitted to the -400 are the PW4056, which produces 56,700lb of thrust at maximum take-off power; the CF6-80C2-B1F, which produces 57,900lb of thrust; and the RB211-524G or H, which produce 58,500 or 60,600lb of thrust respectively.

The engines are all similar; they are all axial flow turbofans with high compression ratios and bypass ratios of around five to one (five times as much air bypasses the engine – ie the hot part – as goes through it). The notable difference between the engines is that the P&W and GE engines have two spools, referred to as N1 and N2, whilst the RR has three, N1, N2 and N3. The other marked difference is that the P&W engine has hydraulically-powered thrust reversers whilst the GE and P&W engines use pneumatic power. The P&W and RR engines use Engine Pressure Ratio (EPR) as a measurement of power, whilst GE uses N1. EPR is the ratio of low pressure turbine outlet pressure (i.e. the rear of the hot part of the engine – the pressure of the exhaust air coming out) to engine inlet pressure (the pressure of the air going in). N1 is the speed of the fan and LP sections of the engine, expressed as a percentage.

From a pilot's perspective the different engines present few issues in operation and it is not a problem to have a 747 fleet with various engine fits (but you cannot intermix the engines on the aircraft). (Hong Kong's Cathay

RIGHT Modular engine pods and engine wing pylons. *(TommyIX/iStock)*

Pacific operates a mixed fleet of RR-powered passenger aircraft and freighters, former Singapore Airlines P&W-powered passenger aircraft, RR and P&W-powered converted freighters, P&W-powered extended range freighters and also GE-powered -8 freighters.)

All three engines are modular. This means that the engine is built in units so that components or sections of the engine can be removed or replaced without having to completely dismantle the engine.

Each engine is attached to a strut by two engine mounts, one near the front and one near the rear of the strut. Close to the top of each strut are three fuse pins, which are designed to shear if the engine seizes, resulting in separation of the engine from the wing rather than separation of the wing from the aircraft.

For the purposes of this book we will take a detailed look at just one engine – that most widely used on the 747-400, which is the General Electric CF6-80C2-B1F.

General Electric CF6-80C2-B1F

The General Electric CF6-80C2-B1F is a dual-rotor axial flow turbofan. It features a single-stage fan, a 4-stage low pressure (LP) compressor, a 14-stage high pressure (HP) compressor, a 2-stage HP turbine and a 5-stage LP turbine. The fan and the LP section are known as the N1 rotor, the HP

LEFT GE CF6-80C2 from the front showing the duct fan and nose cone. *(Jonathan Falconer)*

BELOW LEFT The bullet fairing houses the front bearing; the big white comma is a visual safety cue for when the engine is turning. *(Jonathan Falconer)*

section is known as the N2 rotor; the two are independent and not connected. In common with the P&W 4000 series engine the N2 rotor drives the accessory gearbox. This engine weighs around 4,450kg.

BELOW View of the back of the fan stage. *(Chris Wood)*

Engine cowls

The engines are mounted in nacelles, which are made up of fixed and hinged cowls. At the front of the engine is a fixed inlet cowl, behind it are a pair of hinged fan cowls that contain access doors for various services as well as a pressure relief door. Next are the two halves of the thrust reverser, the cold fan air exhausts at the rear of them. Around the turbine section are hinged core cowl panels and at the rear is an exhaust sleeve. This is where the hot turbine air exhausts. The hinged cowls are attached to the strut, hinged at the top and latched at the bottom.

Engine indicating system

Each engine has a full set of digital instruments in the flight deck. N1 rotational speed and Exhaust Gas Temperature (EGT) are permanently displayed on the upper EICAS screen. N2 rotational speed, fuel flow, oil pressure, oil temperature, oil quantity and engine vibration are on the secondary indications, the engine synoptic, which is selectable on the lower EICAS screen. N1, N2 are shown as percentages: 100% N1 is 3,280rpm and 100% N2 is 9,827rpm (the maximums are 117.5% for N1, which is 3,854rpm, and 112.5% for N2, which is 11,055rpm).

The secondary indications display when the screens initially receive electrical power, when selected, and when any engine fuel control switch is moved to CUT OFF.

All the indications have a digital display and, with the exception of fuel flow and oil quantity, all the indications also have a moving vertical strip indicator. These are colour-coded, with the normal operating range displaying in white, the caution range in amber (EGT, oil temperature and pressure only) and the limit in red (EGT, N1, N2, oil temperature and pressure). If a limit is exceeded, the indication turns red and then, even if the indication reduces below the limit, for N1, N2 and EGT the digital display is surrounded by a box which remains red. If a secondary indication limit is exceeded 'VVVVV' displays in the status prompt, and the secondary indications appear in a partial format at the bottom of the primary EICAS.

N1 and N2 information is derived from tachometers that supply speed signals to the Electronic Engine Control (EEC) system, to the Airborne Vibration Monitoring (AVM) system and directly to the EIUs as a back-up. The EECs send the data to the EIUs for display on the EICAS.

EGT information is derived from eight EGT probes at the LP turbine inlet section of the engine. The information is sent to the EECs, which process it and send it to the EICAS for display.

Fuel flow information is taken from a fuel flow transmitter, which is downstream of the engine fuel system Hydro Mechanical Unit (HMU). It is also fed to the Flight Management System (FMS) for fuel usage calculations.

Indications of oil pressure, temperature and quantity are generated by the oil indication sub-system.

The AVM system uses accelerometers in the fan and core sections of the engines to detect vibration. The system processes vibration and speed signals and computes a vibration value for EICAS to display. This will be: fan, LP turbine or N2 vibration, whichever has the higher value being the signal displayed. If the system cannot detect the source of the vibration it displays a BroadBand (BB) figure. However, all the values can be seen on the EICAS maintenance page, accessible via the CMCS.

Engine control system

Engine power output is controlled by Electronic Engine Controls (EECs), with each engine having its own EEC mounted on the left-hand side of the fan case. Each EEC is powered by its own alternator, mounted on the front left side of the accessory gearbox. This means that in the event of the loss of the aircraft's electrical power, engine control is still available. This is important as there is no direct mechanical connection between the thrust levers and the engines.

The EEC controls fuel flow, engine airflow, engine compartment cooling airflow, ignition, the engine start valve and the engine autostart system. It has two channels, Normal and Alternate, and either channel is capable of controlling the engine. If the Normal channel fails, the EEC automatically switches to the Alternate channel and it can also be selected manually.

The position of the thrust levers determines the desired power, expressed as N1. The EEC calculates the correct N1, compares it with the actual N1 and adjusts the fuel flow through the fuel metering valve in the engine fuel system Hydro Mechanical Unit (HMU). The EEC corrects for changes in outside air temperature and pressure, and also for the effects of any air bleeds from the engine, so that in a climb there is no need to move the thrust levers to maintain the power output.

The EECs constantly compute the maximum allowed N1 and they will limit thrust to this figure, regardless of the position of the thrust levers. They also provide N1 and N2 overspeed protection. The EECs send engine data and fault information to the EIUs for display on the EICAS system, and recording on the CMCS.

There are two idle power settings – Minimum Idle and Approach Idle. Approach Idle is automatically selected in flight by the EEC when particular selections are made:

■ Trailing edge flaps at a landing setting.
■ Engine continuous ignition selected ON.
■ Engine nacelle thermal anti-ice selected ON.

This provides a higher idle power, which reduces the risk of flameout in icing conditions and allows faster engine acceleration during the landing phase, which is useful in a go-around situation.

The Alternate mode works in the same manner as Normal mode, but does not compensate for environmental changes. It also does not provide maximum N1 protection, so

LEFT Electronic Engine Control and associated cabling on the left-hand side of the fan case. *(Chris Wood)*

Alternate mode should not be selected with the engines at high power settings, to prevent over boosting the engines. Care is required when moving the thrust levers fully forward when in Alternate mode, for the same reason.

Engine fuel system

Fuel for each engine is normally pumped under pressure from the tanks through the engine's tank-mounted spar valve. Next it passes through a two-stage engine-driven fuel pump, which

LEFT Fuel injectors and igniter (at the end of the cable). *(Chris Wood)*

increases its pressure, then through a fuel/oil heat exchanger. This serves two functions: it uses warm engine oil to warm up the cold fuel and uses the cold fuel to cool the hot engine oil. Next it passes through a filter, where the flow is divided into high-pressure fuel fed directly to the Hydro Mechanical Unit (HMU) and servo fuel, which is heated before reaching the HMU. When the fuel control switch is moved to RUN, if there is sufficient fuel pressure, the Pressure Regulating and Shut-Off Valve (PRSOV) in the HMU opens. The fuel then passes through a fuel flow transmitter, an IDG fuel/oil heat exchanger and into a fuel manifold on its way to the nozzles in the combustion chamber of the engine. Any excess fuel at the HMU is returned to the engine-driven fuel pump.

Servo fuel is used to operate the fuel metering valve in the HMU and to operate variable stator vanes and variable bypass valves to control the airflow through the compressors, all managed by the EECs.

Engine air system

The engine air system has two functions: the cooling of the engine and its accessories, and compressor airflow control to maintain compressor efficiency and prevent surges. Cooling is achieved using air bled from various sections of the engine through externally mounted valves; compressor airflow control is achieved by using variable stator vanes (VSVs) and variable bypass valves (VBVs).

Fan air is used to cool the accessories, controlled by the core compartment cooling valve (CCCV), which is mounted on the left side of the engine.

LP compressor air is used to cool the centre bore of the engine, controlled by three bore cooling valves (BCVs).

HP compressor air is bled through two eleventh-stage cooling valves (ESCVs) into manifolds for cooling the HP turbine second-stage blades and nozzles. They are mounted on either side of the engine.

The turbine casing is cooled to reduce expansion and therefore reduce turbine blade tip clearance. Two valves, one for the HP turbine, the other for the LP turbine, bleed air into a pair of manifolds that encircle their respective casings.

All the valves are controlled by the EECs, with all of them except the CCCV providing feedback.

The 12 VBVs maintain optimum airflow into the HP compressor by allowing some air to bypass the LP compressor. The VSVs maintain optimum airflow through the HP compressor by varying their angle to the airflow. They are both controlled by the EEC using actuators powered by servo fuel from the HMU.

Engine oil system

The engine has an internal oil system that lubricates, cools and cleans the bearings and gearboxes. The oil is stored in a tank mounted on the right-hand side of the fan casing, and gravity-fed to a pump, driven by the accessory gearbox. From the pump it passes through a filter, which has a bypass that opens if the filter becomes blocked. Filtered oil is then supplied to the bearings and the two gearboxes (accessory and transfer).

The oil and any contaminants are then removed from the bearing compartments and gearboxes by a scavenge system, which consists of 5 scavenge pumps – 3 for the bearing compartments and 2 for the gearboxes. The scavenge pumps and the pressure pump are all part of the same assembly, driven by the accessory gearbox. The scavenged oil is passed through a magnetic chip detector, which captures any magnetic particles as a way

BELOW Final stage of the LP turbine. The fairing houses the tail bearing and it shapes the nozzle to optimise thrust. *(Chris Wood)*

of monitoring for excessive wear, to a servo fuel heater and the fuel/oil heat exchanger, both of which have bypasses. These use warm engine oil to warm up cold fuel, and also to cool the oil. The oil then passes through a scavenge filter, which also has a bypass, on its way back to the tank.

There is an oil breather system that vents bearing seal pressurisation air. It also pressurises the scavenge pump inlets to assist scavenging.

Oil temperature, pressure and quantity are displayed on the engine synoptic on the lower EICAS. EICAS messages additionally warn of low oil pressure, high oil temperature and provide warning of the impending opening of an oil filter bypass.

Engine starting system

The start system supplies air through a starter to initiate N2 rotation for engine start. If an in-flight engine restart is required, the starter can be utilised to assist a windmilling engine.

Engine starting is normally done automatically using the autostart system, if fitted, which utilises the EEC to sequence the start. After initiating rotation, fuel is introduced, ignited, the engine accelerated to idle speed and then stabilised.

During the start the EEC monitors the

ENGINE START SEQUENCE

- Pull out the Start switch in the overhead panel (this opens the engine bleed valve and arms the start valve).
- Move the fuel control switch on the centre console to RUN (this initiates the start allowing the EEC to open the start valve).
- Bleed air now flows to the starter motor, which is connected to the N2 rotor, which will start rotation of the engine.
- At a predetermined N2 speed (around 15% – shown on the secondary EICAS N2 indication as the fuel on indicator), fuel is introduced via the fuel metering valve in the HMU and the igniters are energised.
- The fuel ignites and the engine accelerates under the control of the EEC.
- At 50% N2 the starter motor cuts out, the start switch returns to the in position and the start valve and engine bleed valve close.

engine parameters and aborts the start if it detects a hot start (high or rapidly rising EGT), a hung start (engine stops accelerating) or no rise in EGT. In these situations it will have three attempts at starting, with ventilation runs between each attempt, before aborting the start sequence if it is still unsuccessful.

It is possible to start the engines without using the autostart system, if it has a fault or is not fitted. In this case the actions are the same but the effects are slightly different. Pulling out the start switch opens both the start valve and the engine bleed valve. This allows bleed air into the engine to start it rotating. When the N2 has accelerated to at least 15%, the fuel control switch is moved to RUN. This opens the spar fuel valve, the fuel metering valve and the engine fuel valve, allowing fuel into the engine, and energises the igniters. The fuel ignites and the engine accelerates to idle power (around 28% N1). As the N2 reaches 50%, the start switch is pushed in to close the start valve and the engine bleed valve. There is no automatic monitoring or controlling for malfunctions so the pilots must abort the start if these occur.

If either the start valve or the pylon valve fails to open they can be opened manually by the ground engineers, using a special tool.

Engine ignition system

The engine ignition system consists of two igniters per engine, which can operate independently or together. They are normally powered by main AC electric power, but can be run by standby AC electrical power. This means that even in the highly unlikely event of all four engines flaming out (and therefore their generators not running) the igniters can still be powered through the standby system (which can receive its current directly from the aircraft's battery.)

The igniters can be selected ON either automatically or manually, using switches in the overhead panel. One igniter is selected ON automatically when an engine's fuel control switch is at RUN, with N2 below 50% (for starting), and also when that engine's nacelle thermal anti-ice is selected ON. The EEC controls the choice of igniter, alternating between the two for each flight. If a flameout is detected, both of that engine's igniters are automatically selected ON, to increase the chances of a relight.

One igniter for each engine is selected ON automatically when the trailing edge flaps are not UP, and manually when continuous ignition is selected ON.

Standby power can be selected manually to the igniters using the standby ignition selector on the overhead panel. This has three positions – 1, NORM and 2. In NORM the ignition system is powered by the main AC power supply if it is available. If it fails then the igniters are automatically powered by the standby AC power supply. If 1 or 2 is selected, then the respective igniters are powered by the standby AC power supply.

Engine thrust reversers

Each engine is fitted with a thrust reverser, which is pneumatically operated and electrically controlled. Interlocks prevent reverse thrust being selected in the air and also on the ground when the thrust levers are not in the closed position. They also require the engine's fire switch to be in its normal position and 28V DC electrical power to be available.

Reverse thrust is selected using levers mounted on top of the thrust levers, initially to the reverse idle detent position. This opens the thrust reverser Pressure Regulating and Shut-Off Valves (PRSOV), which are in the struts, and Directional Pilot Valves (DPV), allowing bleed air to the reverse thrust actuators. Amber REV unlocked lights will illuminate on the upper EICAS above the N1 indication.

The actuators move a pair of sliding sleeves on the sides of each of the engines, which are attached by drag links to sets of six blocker doors on either side. These blocker doors prevent the flow of fan exhaust air through the engines and divert it to cascades that redirect it forwards. When both sleeves are fully deployed the REV lights turn green and reverse thrust can be increased above idle, the amount of power being controlled by the position of the reverse thrust levers through the EECs.

To cancel reverse thrust the levers are returned to the stow position. This opens the PRSOVs and, as the DPVs are spring-loaded to close, this allows bleed air to the stow side of the actuators. As the sleeves move forward the blocker doors move back into the sleeves. The EEC prevents movement of the thrust levers until the sleeves are fully stowed. Only cold fan air is used for reverse thrust.

Pratt & Whitney PW4056

The first 747-400 to fly was fitted with the P&W engine. Like the CF-6 this is a dual-rotor axial flow turbofan. It has a single-stage fan, a 4-stage low-pressure (LP) compressor, an 11-stage high-pressure (HP) compressor, a 2-stage HP turbine and a 4-stage LP turbine. The fan and the LP section are known as the N1 rotor, the HP section is known as the N2 rotor; the two are independent and not connected.

Hot exhaust gases from the combustion section drive the turbines. The LP turbine drives the fan and the LP compressor (N1), the HP turbine drives the HP compressor (N2). The N2 rotor also drives the accessory gearbox via an angle drive gearbox. This drives a combined oil pressure and scavenge pump, a two-stage engine fuel pump, an engine-driven hydraulic pump, an Electronic Engine Control (EEC) generator and an Integrated Drive Generator (IDG). The N2 rotor is also attached to a starter motor for engine starting.

With this engine 100% N1 is 3,900rpm and 100% N2 is 9,905rpm (the maximums are 111.4% for N1, which is 4,012rpm, and 105.5 % for N2, which is 10,450rpm).

Only 20% of the thrust comes from the

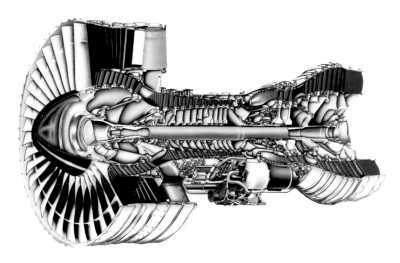

hot section of the engine, the remaining 80% comes from the cold, fan-driven air that bypasses the hot section. Each P&W engine weighs around 4,216kg.

Rolls-Royce RB211-524G/H

The Rolls-Royce RB211-524 is a triple-rotor axial flow turbofan. The triple-rotor design allows the engine to have a shorter, stiffer structure, and also enables the rotors to run nearer their optimum speeds, reducing the need for compressor variable guide vanes. It has a single-stage fan, a 7-stage Intermediate Pressure (IP) compressor, a 6-stage HP compressor, a single-stage HP turbine, a single-stage IP turbine and a 3-stage LP turbine. The fan and LP turbine comprise the N1 rotor, the IP section is the N2 rotor and the HP section is the N3 rotor; the three are independent and not connected. The N3 rotor drives the engine accessory gearbox.

For this engine 100% N1 is 3,900rpm and 100%, N2 is 7,000rpm and 100%, N3 is

ABOVE Cutaway illustration of the Pratt & Whitney PW4000. *(Pratt & Whitney)*

BELOW A ground engineer inspects inside the cowls on a Rolls-Royce RB211-524G or H on a 747-400. *(Rolls-Royce Plc)*

10,611rpm (the maximums are 110.5% for N1 which is 4,310rpm, 110.5% for N2 which is 7,539rpm, and 99.2% for N3 which is 10,526rpm).

With this engine the hot air creates around 21% of the thrust; the remaining 79% comes from the cold fan air. However, these two airflows are mixed and pass out through an integrated exhaust nozzle. The RB211 is much heavier than the other engine options. Each engine weighs approximately 5,688kg (-524G) or 5,790kg (-524H).

An upgrade for the -524G and H model engines became available in 1997. This involved replacing the HP section of the engine with one based on the HP section of the Trent 700 engine. The benefits of this modification were a 2% lower fuel burn, a lighter engine, lower emissions, better performance and lower maintenance costs. Modified engines are referred to as -524G-T or -524H-T models.

Auxiliary Power Unit (APU)

Like all modern jet airliners the 747 has an extra engine. This is known as the Auxiliary Power Unit (APU) and the one fitted to the 747-400 is the Pratt &Whitney Canada (P&WC) PW 901A. It is mounted in the extreme rear of the fuselage, outside the pressure hull behind a titanium firewall. Its purpose is to supply electrical power and an air supply to the aircraft when it is on the ground, and in some circumstances to supply air when airborne.

The PW 901A is a centrifugal flow gas turbine based on P&WC's JT-15D turbofan, but with four stages rather than the six of the JT-15D. It has two counter-rotating single-stage turbines, a low-pressure power turbine and a high-pressure turbine. Each turbine is connected by a shaft to a single-stage impeller, with the power turbine driving the load compressor impeller, shown as APU N1 on the EICAS, and the high-pressure turbine driving the high-pressure impeller, which is shown as N2. It has an annular reverse flow combuster with 14 fuel nozzles and dual high-energy igniters.

Operation of the APU is managed by an APU controller (APUC), which is a Full Authority Digital Engine Control (FADEC) system. This controls the start sequence, regulates the fuel flow to maintain a constant speed through the fuel metering unit (FMU) and monitors all the operating parameters. If it detects a malfunction this information is sent to the EICAS and CMCs. If it is a serious malfunction it will also initiate an automatic shutdown of the APU.

Mounted on the right-hand side at the rear of the fuselage is the APU door. This is closed when the APU is not running, but opens to 45° when the APU is selected ON. It has a third position, 15° open, which is selected if the APU is running when the aircraft is airborne. Door movement is powered by an electric actuator.

Air is drawn in past the door and is directed radially to both the high-pressure impeller, where it is compressed and passed to the combustion chamber, and to the load compressor impeller. Air in the combustion chamber is mixed with fuel and ignited, with the hot gas driving both turbines before exiting out of the exhaust duct. The load compressor impellor supplies air to

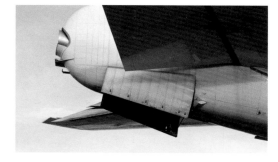

the pneumatic manifold though a bleed air valve, mounted in the left-hand body gear wheel well, and a non-return valve. The amount of air supplied is controlled by variable inlet guide vanes, which are also used to manage air flow during the APU start sequence. Inlet guide vane position is controlled by the APUC.

APU bleed air is used primarily for engine starting and conditioning the cabin, but as it is fed into the pneumatic manifold it can also be used to power all the pneumatic systems.

The load compressor shaft drives a load gearbox which drives two 90 KVA electrical generators and a cooling fan. The cooling fan provides air to cool the generators, the APU oil via a heat exchanger and the APU compartment.

There is an accessory gearbox, driven from the N2 rotor, which drives a fuel pump, an oil supply pump and three oil scavenge pumps. It is also connected to the APU's electrical starter motor.

The APU has an integral oil system which comprises a 20.8-litre tank, a pressure pump, three scavenge pumps, a filter, a pressure regulating valve and an air-cooled oil cooler. There are also two magnetic chip detectors, one in each gearbox.

Fuel for the APU is drawn from the No 2 main fuel tank through a fuel shut-off valve mounted on the rear spar, and a shrouded fuel line. There are two electrical fuel pumps, a dc one powered by the APU battery, and an ac which is operated whenever there is an ac electrical supply to power it.

There is a three-position APU master switch in the flight deck: OFF, ON and START (which is spring-loaded back to ON). When switched to the ON position, either the dc or ac fuel pump runs (depending on electrical power supply), the spar fuel valve opens and the APU door opens. When switched momentarily to START several things happen: the APUC is powered and performs a self-test, the primary fuel valve in the FMU is moved to the RUN position, the secondary fuel valve is readied for start, the load compressor inlet guide vanes move to the START position and the starter motor and igniters are energised.

At 8% N2 the secondary fuel valve in the FMU moves to the RUN position, allowing fuel into the combustion chamber for ignition. At 40% N2 the starter motor cuts out and the

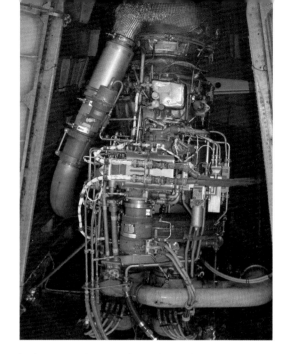

igniters are de-energised. At 55% N2 the inlet guide vanes move to the NO LOAD position and a timer for the low oil pressure switch starts. If sufficient oil pressure is not detected after 15sec the APU is automatically shut down. Two seconds after reaching 95% N1, air and electrical power are available and an APU RUNNING message appears on the EICAS.

The APU generators will then be available (shown by AVAIL lights in the switches) and the APU bleed valve will be open providing bleed air to the pneumatic manifold. To switch on the APU generators press the switches; the ON lights in the switches will now illuminate and the AC ground electrical supply (if connected) will be disconnected. The APU stabilises at 100% N1.

To switch off the APU move the master switch to OFF. This will disconnect the APU generators and close the bleed air valve. The APU will run for a further minute as a cooling down period, prior to shutting down. The fuel valve will be closed, the fuel pump switched off and the APU door shut.

The APU is designed to be left unattended so it has an automatic shutdown system in case any significant malfunctions are detected. If a fire is sensed the APUC will shut down the APU, activate the fire warning horn and illuminate the APU fire switch on the flight deck overhead panel. Pulling the switch will arm the fire extinguisher, which is fired by rotating the fire switch. There is also a fire switch and means of firing the fire extinguisher in the right-hand body gear wheel well.

Chapter Four

Systems

⊶━━━━━━━━━━━━━━━━━━━━━━━

The 747-400 is an electronic
aircraft, and a highly complex one
at that, in which every operating
system is interrelated. At the heart
of the aircraft is the Indicating
System, to which everything
is connected. It is through the
actions of the Indicating System
that the aircraft 'talks' to the pilots
and the engineers.

**OPPOSITE The electronic flight deck environment of the
Boeing 747-400.** *(Thierry Deutsch)*

Apart from the engines where there is a choice of three different options, most 747-400s are very similar, far more so than the 'Classic' 747s ever were. There are some minor variations in fit, so in the descriptions that follow there may be a few small differences between what is described and what is fitted to any particular aircraft that the reader may be familiar with.

Indicating systems

Before taking a detailed look at the various systems, it is worth reviewing the aircraft's indicating and warning systems. These represent a radical change and improvement over the systems fitted to the 'Classic' 747s.

The aircraft is equipped with an Integrated Display System (IDS – often referred to as a 'Glass Cockpit') which is made up of an Electronic Flight Information System (EFIS – pronounced 'e-fis') and an Engine Indication and Crew Alerting System (EICAS – pronounced 'i-cas'). The information from these systems is displayed on six screens, referred to as Integrated Display Units (IDUs). Each pilot

has two IDUs in front of them showing EFIS data; the outer screen is known as the Primary Flight Display (PFD), the inner screen as the Navigation Display (ND). The other two IDUs are used to present EICAS information. The IDUs are all identical and are interchangeable.

The two systems communicate electronically through a device called an EFIS/EICAS Interface Unit (EIU). There are three EIUs and they are all located in the MEC. One is in 'control' with the other two as back-up in case the controlling one fails. All the aircraft systems are continuously monitored by all three EIUs; they provide information to the EICAS displays and generate EICAS messages as a crew alerting function if they detect any failures.

There are two EICAS screens, an upper and a lower one. The upper screen shows permanently and displays the main engine instruments, EICAS messages and some other useful pieces of information. The lower screen is selected on at the EICAS select panel and can display information on the various aircraft systems. These displays are known as system synoptics, and the systems can be viewed one

RIGHT The flight deck of a Japan Airlines (JAL) 747-400. *(Norio Nakayama/Wikimedia Commons)*

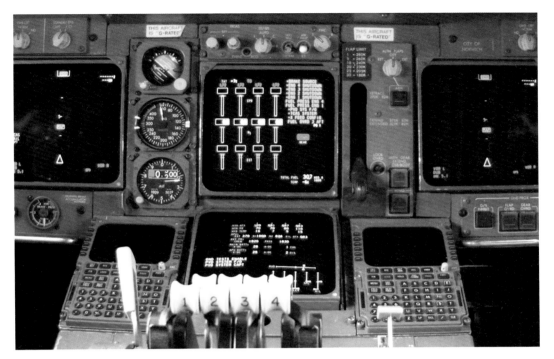

at a time. There are eight synoptics that can be selected:

- Engine (ENG)
- Status (STAT)
- Fuel (FUEL)
- Electrics (ELEC)
- Environmental Control System (ECS)
- Hydraulics (HYD)
- Doors (DRS)
- Landing gear (GEAR)

They are selected on by pressing the associated button on the EICAS select panel. If the same button is pressed again the display clears and the screen goes blank; if a button for a different system is pressed the associated synoptic displays.

The lower EICAS can also be used to show information from the Central Maintenance Computer System (CMCS).

If the screens fail there are some automatic changeovers. If the outer screen fails the PFD information is automatically transferred to the inner, ND, screen (PFD information has priority over ND information). If the upper EICAS screen fails the information is automatically transferred to the lower EICAS. It is also possible to change screens manually using switches on the pilots' main panels.

There are three types of message that can be generated:

- Alert message
- Memo message
- Status message

Alert messages are further broken down into three categories:

- Warning. These are the highest priority (so will always be at the top of the screen), appear in red and cannot be cancelled. The only way to remove this type of message is to fix the problem that is causing it. For example, if it is a fire warning, put out the fire!
- Caution. These are the next highest priority, appear in amber and can be cancelled, even if the fault is still present.
- Advisory. These are the lowest priority, also appear in amber but are indented one space, and can be cancelled.

The caution and advisory messages can be cleared from the EICAS screen by pressing the cancel button on the EICAS control panel. However, they can be retrieved, if they are still active, by pressing the recall button on the EICAS control panel.

Associated with the EICAS messages are

two master caution lights fitted on the combing in front of the pilots. These act as attention getters when EICAS messages are posted.

In certain circumstances, such as during engine start and during take-off, some EICAS messages are automatically inhibited.

Each Alert message has an associated checklist in the Quick Reference Handbook (QRH). In the event of a system failure, an EICAS message will appear on the centre EICAS screen and all that is required is to pick up the QRH, turn to the relevant page for that message and follow the procedure. If that sounds very simple, it is! However, it gets interesting when dealing with multiple failures (which are very rare occurrences, except in the simulator!); the aeroplane will advise you of the failures (you get a selection of EICAS messages) but it cannot decide a priority for dealing with them; that is still a job for the pilots.

Memo messages appear in white and are reminders to the crew, for example that certain systems are selected on that are not used all the time, such as anti-icing or the parking brake.

Status messages appear on the lower EICAS screen and can indicate a lower level of failure, although many, if not all, EICAS Alert messages have associated Status messages. If a Status message is generated a STATUS prompt appears on the upper EICAS. Status messages remain displayed and are cleared through the CMC after the flight as a maintenance function.

Central Maintenance Computer System

Virtually the entire aeroplane is monitored by the Central Maintenance Computer System (CMCS). Signals from the Integrated Display System (IDS) are sent to the CMCS via the EIUs and it collects, stores and displays information on any faults it detects. There are actually two Central Maintenance Computers (CMCs) fitted in the Main Equipment Centre (MEC). They can generate around 7,800 different fault messages and store up to 500 faults per flight. Whilst this is primarily a maintenance function, this information can be retrieved in flight, although it is inhibited when the aircraft's groundspeed is above 50kt and its altitude below 10,200ft.

The CMCS also has a test function that can be used to check the integrity of the aircraft systems which have a Built In Test Equipment (BITE) function. When selected the CMCS will run the desired test and display the results. This function is only available on the ground.

The aircraft can be fitted with a printer on the flight deck, and information from the CMCS can be printed. There are also certain failures or events, such as a heavy landing or an engine limit exceedence, which will automatically generate a printed report.

CMCS reports can also be transmitted via the aircraft's Aircraft Communications And Reporting System (ACARS), and it is possible for it to be interrogated from the ground.

CMCS fault codes are listed in the engineer's Fault Isolation Manual (FIM). This will detail the flight deck effects of the fault, the corrective action to be taken to fix the fault and the reference in the Aircraft Maintenance Manual (AMM) for the procedure.

Flight instruments

Information for the flight instrument displays is obtained from two (optionally three) air data computers (ADCs). The ADCs get their information from four pitot static probes, two total air temperature probes, two angle of attack probes and four static ports. The ADCs process the information and produce speed, altitude, temperature, pressure and angle of attack data, which is fed to the aircraft systems that need this information, such as the Flight Management Computers (FMCs), Electronic Engine Controls (EECs), Inertial Reference Systems (IRS), Flight Control Computers (FCCs), the Stall Warning Computers and the ATC transponder.

All instrumentation is displayed electronically

across six EFIS screens. There are two main displays for each pilot, the Primary Flight Display (PFD) which shows the flight instruments and the Navigation Display (ND) which presents navigation information. The other two screens are used by the EICAS to show engine and systems information. All the displays have brightness controls, and they also have sensors to measure ambient light and automatically adjust the brightness.

The PFD has a horizon display with a flight director. It also shows altitude and indicated airspeed or Mach number, in a tape format. In each case both the desired value, as selected by the FMC or on the autopilot flight director system (AFDS) controller, and the actual value (from the ADC) are shown. There is also a rate of climb or descent indicator and a compass display showing the aircraft's heading. Autopilot and Flight Director modes are annunciated at the top of the PFD. Radio altimeter reading is shown at the top on the right-hand side, and the barometric pressure setting is shown at the bottom. ILS indications are also displayed on the PFD.

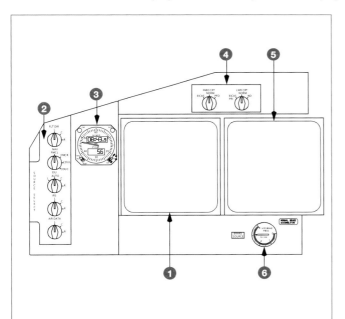

Left forward instrument panel, Captain.

1 Primary Flight Display (PFD)
2 Source select switches
3 Clock
4 CRT selector switches
5 Navigation Display (ND)
6 Brake accumulator pressure gauge

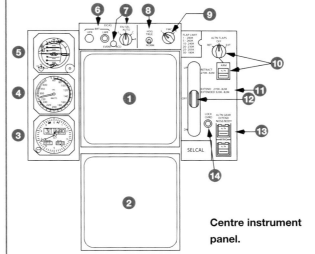

Centre instrument panel.

1 Upper Engine Indicating and Crew Alerting System (EICAS)
2 Lower EICAS
3 Standby Altimeter
4 Standby Airspeed Indicator
5 Standby Attitude Indicator
6 Brightness controls
7 EIU selector switch
8 Compass true or magnetic selector
9 FMC selector
10 Alternate flap switches
11 Landing gear panel
12 Landing gear lever
13 Alternate landing gear extension switches
14 Landing gear lock override

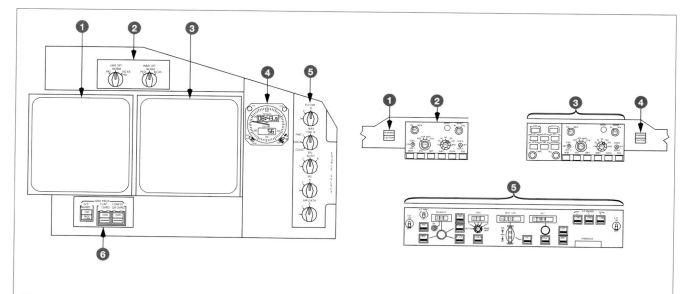

Right forward instrument panel, First Officer.

1. Navigation Display (ND)
2. CRT selector switches
3. Primary Flight Display (PFD)
4. Clock
5. Source selector switches
6. GPWS switches

Glare shield panel

1. Master warning and caution light
2. Captain's EFIS control panel
3. EICAS control panel and First
4. Officer's EFIS control panel
5. Master warning and caution light
6. AFDS control panel

The airspeed display also shows take-off and landing reference speeds, as generated by the Flight Management Computer (FMC) and/or programmed through the Control Display Unit (CDU), flap manoeuvring speeds (minimum speed for the selected flap setting), maximum speeds (as limited by landing gear, flap or maximum operating speed, known as VMO or MMO) and minimum manoeuvring speeds.

The maximum and minimum manoeuvring speeds show as a yellow band on the speed tape; the maximum and minimum speeds show as red bricks on the speed tape.

There is a radio altimeter system for use at low altitudes, below 2,500ft. There are three separate radio altimeters, each with their own antennas and transmitter/receivers: left, centre and right. Radio altimeter information is also displayed on the PFDs.

Put simply, the PFD shows all the information required to actually fly the aircraft (although that is not quite true as it does not show engine power, which is also quite useful to know!).

The ND has several different modes – APP (approach), VOR, MAP and PLAN, with MAP mode being used for most of the time. In this mode the aircraft's heading or track (another option) is at the top of the screen and the aircraft's planned track, as loaded into the FMC, is displayed as a pink line (officially its magenta!) with all the route waypoints. In this mode it can also have weather radar or terrain returns and Traffic Alert and Collision Avoidance System (TCAS) returns superimposed on the map. These can be selected using switches on the EFIS control panel. When weather radar is selected a variable range scale with increments at 5, 10, 20, 40, 80, 160, 320 and 640 nautical miles displays. Additional switches allow the pilots to display information such as airfields, navigation beacons and waypoints other than the ones on the route.

In PLAN mode North is displayed at the top and again the route, with waypoints, is shown in magenta.

In APP mode heading is displayed at the top, and instrument landing system (ILS) information is shown, such as the selected inbound course, and localiser and glideslope indications. Also on display are speeds (both groundspeed and airspeed), wind and basic ILS data.

The VOR mode is very similar to the APP mode; heading is displayed at the top and information from the selected VOR radio navigation beacon is shown.

The MAP, APP and VOR modes can show either a full compass rose or a partial one (known as the expanded mode). Selections between the two are made by the CTR (centre) switch on the EFIS control panel.

The aircraft has a set of standby instruments comprising a magnetic compass, on the centre post above the glare shield, a standby attitude indicator, a standby airspeed indicator and a standby altimeter, all fitted to the right of the Captain's PFD. The attitude indicator is a mechanical gyro, and is powered directly from the aircraft battery. The airspeed indicator and altimeter are traditional round dial instruments and receive uncorrected pitot static data. In later aircraft all three were replaced by a single electronic instrument called an Integrated Standby Flight Display (ISFD). The standby attitude indicator and the ISFD can also display ILS information. This means that if all other information is lost the aircraft can still be flown on instruments up to and including an ILS approach.

Navigation

Primary navigation is achieved using an Inertial Reference System (IRS), which consists of three Inertial Reference Units (IRUs), containing laser ring gyros, which are in the MEC. These are refined versions of the Inertial Navigation Systems (INSs) fitted to the earlier 747s. This type of system was originally designed for the Apollo moon programme – Apollo 11 had one system and got to the moon and back; the 747 has three!

Each IRU has a four-position switch on the pilot's overhead panel, marked OFF, ALIGN, NAV (navigation) and ATT (attitude). IRS alignment is initiated by turning the switches straight to NAV and entering the aircraft's position (latitude and longitude) through one of the control display units (CDUs). The IRUs take 10min to align and the aircraft must not be moved during this period. Once they are aligned they provide position, attitude and heading information. Combined with outputs from the ADCs, data such as groundspeed and wind information can be calculated.

IRU signals are supplied to systems such as the Flight Management Computers (FMCs), Flight Control Computers (FCCs), Control Display Units (CDUs), the yaw dampers and the autobrakes.

The IRUs can be realigned on the ground by switching to ALIGN, entering the aircraft's position and switching back to NAV. This method of alignment only takes 30sec.

The ATT mode provides a means for the IRS to provide attitude data if its navigation function fails. Selecting ATT results in the IRS entering the align mode for 30sec (so it should be done with the aircraft straight and level). It will then provide an attitude display to the PFD.

The aircraft can be fitted with a Global Positioning System (GPS), comprising two independent receivers. Its primary function is to update the IRS position within the FMC; however, GPS position can also be displayed on the CDU and on the ND.

The aircraft has twin VOR/DME receivers, twin ADF receivers and triple ILS receivers. The VOR/DME and ILS can be tuned automatically by the FMCs or manually through the CDUs on the NAV/RAD page. The VOR/DME is also used to update the IRS. The ADF receivers can only be tuned manually, through the CDUs. Indications from the radio navigation aids are displayed on the PFD and the ND.

There is a weather radar system that can be used to detect, and hence avoid, severe weather. It is a dual system using a single antenna in the nose. Weather radar returns are displayed on the pilots' NDs when WXR is selected on the EFIS control panel, in all modes except PLAN, plus VOR and APP when they are in centre mode. The returns are colour coded, with the strongest being red, next strongest yellow and then green. Red areas are to be avoided! The radar is controlled using switches on the weather radar control panel at the front of the centre console. The antenna can be tilted and it is stabilised by the IRS system.

There is a dual Air Traffic Control (ATC) system containing two mode S transponders. The transponder control panel, on the right-hand side of the centre console, also manages the Traffic Alert and Collision Avoidance System (TCAS).

Aural warning system

Certain failures and conditions will generate aural and visual warnings. These are provided by the Aural Warning System, which receives signals from numerous sensors in the aircraft systems. The signals are processed in the

Modular Avionics Warning Electronics Assembly (MAWEA), located in the MEC. The aural warnings can be tones or voice messages, the visual warnings are displayed on various panels, such as the PFDs and the NDs.

Traffic Alert and Collision Avoidance System (TCAS)

The aircraft is fitted with a TCAS (pronounced *tee-cas*) II system. This uses its own mode-S datalink function and other aircraft's ATC transponders and has two functions. It detects other aircraft within a defined area and displays their position and relative height information on the pilots' NDs. If other aircraft get too close it generates instructions to avoid a collision.

Traffic detected by the system displays as a diamond with a number. The number indicates the other aircraft's relative difference in altitude: for example +10 means it is 1,000ft above, -7 means it is 700ft below. If it is climbing there will be an arrow pointing upwards, if descending an arrow will point down.

There are four types of traffic symbols that can be displayed:

- Other traffic – this shows as a hollow white diamond.
- Proximate traffic – this is a filled white diamond (it is getting closer).
- Traffic Advisory (TA) – the filled diamond turns yellow and the system announces 'Traffic, traffic!' (getting close enough to be of concern).
- Resolution Advisory (RA) – the yellow diamond turns red (it is too close, do something NOW!).

When an RA is generated, in co-ordination with the other aircraft's system, climb or descent instructions are given to avoid a collision. These are displayed on the pilots' PFDs and complemented by voice warnings, such as 'Climb, climb!', or 'Descend, descend!' When the danger has passed the PFD warnings clear and the system announces 'clear of conflict'. Obviously, for the system to work both aircraft must be fitted with functioning transponders with an altitude mode!

The system is controlled by a TCAS computer located in the MEC.

Ground Proximity Warning System (GPWS)

The aircraft was originally fitted with a GPWS, but an Enhanced GPWS (EGPWS) became available in the mid-1990s. The GPWS uses inputs from the radio altimeter and various other aircraft systems to provide warning of impending ground contact by generating audio and visual warnings. It is active between 2,450ft and 30ft radio altitude. The EGPWS includes a terrain database and so can give warning of terrain ahead of the aircraft. Warnings provided include excessive descent rate 'Sink Rate!', excessive terrain closure rate 'Terrain!', excessive descent after take-off 'Don't sink!', unsafe terrain clearance without the landing gear extended 'Too low gear!' and deviation below the ILS Glideslope 'Glideslope!'

The GPWS detects windshear and generates visual and audio 'Windshear!' warnings. The Predictive Windshear System (PWS) uses the aircraft's weather radar system to detect windshear ahead of the aircraft, generating a 'Windshear ahead!' warning.

The GPWS also provides voice altitude call-outs during descent and approach.

Flight Management Computer System (FMCS)

A whole book could be filled about the FMCS functions. It is by far the biggest chapter in the pilot's manual, so this is a very simplified look at a very complex area.

The aircraft has a Flight Management Computer System (FMCS), consisting of two Flight Management Computers (FMCs). Normally one works as master, the other as slave, poised to take over if the master fails. They contain a database of navigational information, such as waypoints, airways, arrival and departure procedures, navigation beacons and airfields as well as performance data for the aircraft. Additional data is entered by the pilots and this is done using three Control Display Units (CDUs). All three can be used to input data to the master FMC, which then copies it to the slave. Normally the left and right CDUs are used for the FMC functions and the centre CDU is used for the ACARS system.

The FMC uses the stored data and the pilot-inputted data to calculate the optimum flight profile, in terms of altitude and speed, known as vertical navigation (VNAV). The route to be flown is programmed by the pilots. This is referred to as lateral navigation (LNAV). VNAV data is displayed on the CDUs, route data is also displayed on the CDUs and on the pilots' NDs.

When the autopilot and autothrottle are engaged, depending on the modes selected, the FMC can be used to control the aircraft's speed. It will also display the optimum altitude, but actual altitude selection is a job for the pilots (as negotiated with ATC).

In the unlikely event of both FMCs failing, the CDUs can be used as an alternate navigation system. They do not have databases, but do store route information from the route being flown (known as the active route). Additional waypoints can be added, using latitude and longitude rather than waypoint names. Radio navigation aids can be tuned using the CDU.

Each pilot has a bank of source selector switches on their forward sidewall. This allows them to select an alternate data source for their instrumentation in the event of failures. With triple systems the aim is to ensure that both pilots always receive information from different sources. Systems for which the pilots have alternate data sources include the flight directors, the FMCs, the IRSs and the ADCs.

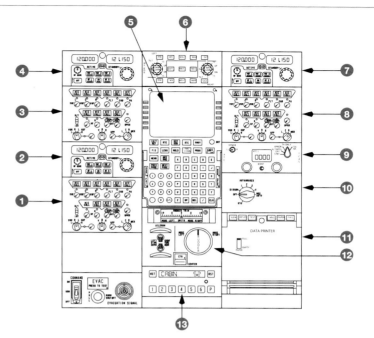

Aft aisle stand and Flight Management Computer.

1   First Observer's radio control panel
2   Centre VHF/HF radio selector panel
3   Captain's radio control panel
4   Left VHF/HF radio selector panel
5   CDU
6   Weather radar controls
7   Right VHF/HF radio selector panel
8   First Officer's radio control panel
9   ATC transponder/TCAS control panel
10 Autobrake selector
11 Data printer
12 Rudder trim knob. This panel also contains the rudder trim indicator, rudder trim centring switch and aileron trim switches
13 Cabin interphone selector panel

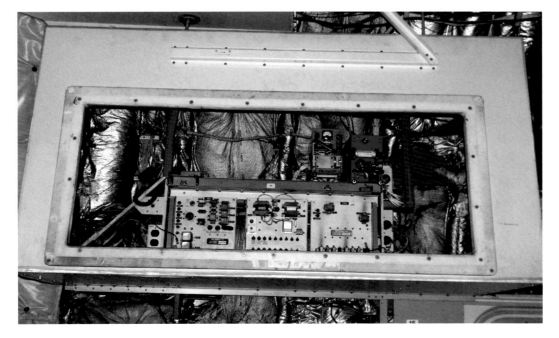

LEFT Cockpit voice recorder, the so-called 'Black Box'. *(Bas Tolsma/ TheDutchAviator.com)*

Communications

The aircraft has a standard fit of three VHF radios (Left, Centre and Right) and two HF radios (Left and Right) for voice communication and two Satellite Communications systems (Left and Right) which can be used for voice or datalink communication. Each pilot position has an audio control panel where radio selections can be made.

The aircraft can also have an Aircraft Communication and Reporting System (ACARS) fitted. This is a datalink system and works through the Centre VHF radio and the Satellite Communications system and is normally controlled through the centre CDU. It can be used to uplink data (such as weather reports), and to downlink data, automatically, by interrogation from the ground or by sending messages (much like texting).

The ACARS system communicates with the Flight Management Computers (FMCs), the Control Display Units (CDUs), the Central Maintenance Computer System (CMCS), the Aircraft Condition Monitoring System (ACMS) and a printer. This means that the engineering department can monitor the state of the aircraft from their offices on the ground.

There are three interphone systems. The flight interphone system allows the pilots to talk to each other and to communicate with the ground crew. The service interphone system provides communication between the pilots, ground crew and engineers, and there are headset jacks all around the aircraft. They can be found in places such as the nose wheel (used by the ground crew during pushback operations), at the refuelling panels, in the MEC and on the engines, to name but a few.

The cabin interphone system provides communication around the cabin and between the cabin and the flight deck. Handsets are installed at all the cabin doors, in the flight deck and at various other locations.

Automatic Flight

The aircraft has an Autopilot Flight Director System (AFDS), consisting of three autopilots and three flight directors, controlled primarily through a Mode Control Panel (MCP) on the pilots' glare shield, and automatic stabiliser trim. The AFDS is made up of three independent systems, with each autopilot being powered by a separate hydraulic system. Each system has a Flight Control Computer (FCC), and three autopilot servo actuators: one each for the elevators, ailerons and rudders.

The flight directors are displayed on the pilots' PFDs. They provide guidance for the pilots to follow the flight profile selected through the FMC at the MCP. Each pilot has an on/off switch for their flight director.

Automatic stabiliser trim works through the autopilot when it is engaged. This operates when the elevators have been deflected for more than 3½sec away from the neutral position, to minimise drag and ensure that the aircraft is trimmed if the autopilot disconnects.

It also has a speed stability trim system, which operates when the autopilot is not engaged, controlled by the stabiliser trim and rudder ratio module (SRM) and occurs in the lower-speed range to improve the aircraft's response to speed changes.

There is an autothrottle system, which can be used throughout the flight, from take-off to landing. It is controlled through the MCP and the CDUs. The MCP has an autothrottle arm switch and two mode switches.

It can be engaged for take-off by pressing either of a pair of Take Off/Go Around (TO/GA) switches, on the front of thrust levers Nos 2 and 3. It is also engaged when certain AFDS modes are selected. It can be released by pressing either of two disengage switches mounted on the outboard side of thrust levers Nos 1 and 4.

In normal operations a flight director is switched on and both an autopilot and the autothrottle are engaged. For an automatic landing all three autopilots engage automatically.

AFDS and Autothrottle modes

The AFDS and autothrottle have several different modes, selected by switches on the MCP. Normally one AFDS vertical (pitch) mode and one horizontal (roll) mode will be in use, with the autothrottle, and the modes in use will be annunciated in green on the top of the pilots' PFDs. Some modes can be armed. These are annunciated in white under the green annunciation.

The roll (lateral) modes are:

■ HDG HOLD: Heading Hold. The aircraft will maintain the heading it was flying when the MCP heading select knob was pushed.

- HDG SEL: Heading Select. The aircraft will fly the heading selected with the MCP heading select knob. With HDG SEL engaged, rotating the knob will turn the aircraft (nearly 400 tons of aircraft being controlled by a finger and thumb!).
- LNAV: Lateral Navigation. The aircraft will follow the route programmed in the FMC. The route is displayed on the ND as a magenta line. If this mode is engaged when the aircraft is not on the programmed route, the mode will arm and a white LNAV annunciation will display. The aircraft will maintain the previously selected lateral mode, such as HDG, and as the aircraft approaches the route it will capture and track it. As it captures the annunciation will change to a green LNAV (referred to as 'LNAV green').
- If LNAV is selected on the ground it will arm and then engage at 50ft above the ground (occasionally it will engage during the take-off roll, if the runway slopes upwards).
- LOC: Localiser. The aircraft will capture and follow the ILS Localiser. This mode can also be armed, when it will annunciate a white LOC. As the ILS localiser is captured the annunciation changes to 'LOC green'.
- ROLLOUT: Rollout. Used during an automatic landing to indicate that the AFDS is tracking the localiser after touchdown. This mode arms at 1,500ft during the approach and activates below 5ft.

The pitch (vertical) modes are:
- ALT HOLD: Altitude Hold. Pressing the altitude selector knob will result in the aircraft levelling off at the altitude at which the knob is pressed.
- ALT: Altitude. In this mode the aircraft is maintaining the altitude selected in the altitude window of the MCP.
- VS: Vertical Speed. Pressing the VS switch opens the MCP vertical speed window. The desired vertical speed, for climb or descent, can be selected using the vertical speed thumbwheel. The speed window will also open and the autothrottle will adjust the power to maintain the selected speed.
- VNAV: Vertical Navigation. In this mode the aircraft will fly a vertical path calculated by the FMC. In a climb the annunciation will be VNAV SPD (Vertical Navigation Speed) as the

ABOVE On the right is the PFD (First Officer's) which displays virtually all the information the pilot needs to fly the aircraft. Left side: Actual Indicated Air Speed (IAS) (289kt), selected Mach No (.84 in magenta), actual Mach No (.844 in white). Centre: Autopilot Flight Director System annunciations: SPD, LNAV, VNAV PTH, a horizon display with Flight Director, partial compass rose display with heading (103° magnetic). Right side: selected altitude (35,000ft in magenta), actual altitude (35,000ft in white), rate of climb or descent (zero), barometric pressure setting for the altimeter (STD, standard setting which is 1013.2 mb or 29.92in hg).

On the left is the ND. The route is shown in magenta, the aircraft is heading towards the next waypoint which is 5620N (N56.00 W020.00) and is estimated to pass it at time 1236.0z. The green dots are weather radar returns. This aircraft displays heading at the top of the ND (as opposed to the option to have the aircraft's track at the top) and does not have the optional range rings with the weather radar. (Thierry Deutsch)

aircraft will be maintaining the FMC calculated climb speed. When the programmed cruising altitude is reached the annunciation will change to VNAV PATH (Vertical Navigation Path). This indicates that the aircraft is following the FMC programmed path.

If the MCP speed window is selected open during the climb, the aircraft will fly the selected speed and VNAV SPD will still be annunciated.

For descent the FMC will calculate a descent path (ideally with idle power) based on any programmed altitude constraints. Whilst the aircraft is following this path the annunciation will be VNAV PATH, and it will endeavour to maintain the FMC calculated or programmed airspeed. However, as it attempts to follow its calculated vertical path, airspeed fluctuations may occur. If it

deviates off the path by more than 500ft, the annunciation will change to VNAV SPD (Vertical Navigation Speed) and the aircraft will now be flying FMC calculated airspeed, but will not be maintaining the path. Likewise if the MCP speed window is selected open, the annunciation will change to VNAV SPD and the aircraft will no longer follow the calculated vertical path.

In the event of an intermediate level-off during either climb or descent the annunciation will change to VNAV ALT (Vertical Navigation Altitude), until a further climb or descent is initiated.

If VNAV is selected prior to take-off the mode will arm, and then engage at 400ft above the ground.

■ FLCH SPD: Flight Level Change, Speed. This mode opens the MCP speed window and can be used for climb or descent. In either case the aircraft will fly towards the altitude chosen in the altitude selector window. For a climb the autothrottle will select climb power; for a descent it will select idle power.

■ APP: Approach. Once locked on to the ILS Localiser, the aircraft will lock on to the ILS Glideslope and follow it. The annunciation is GS (Glideslope) and this mode is armed (annunciated in white) until the glideslope is captured, when it changes to 'GS green'.

■ FLARE: Flare. Like ROLLOUT it is used during an automatic landing and is armed at 1,500ft. It activates at around 50ft to indicate that the autopilot is flaring the aircraft for landing, and although it deactivates at touchdown it is used to smoothly lower the nose wheel.

The autopilots can be used from a height of 250ft after take-off until coming to a stop on the runway after landing. Normally only one autopilot

is engaged, and it only operates in the pitch and roll channels. However, during an automatic ILS approach, as the aircraft passes through 1,500ft all three autopilots engage. When this occurs the rudder channels engage. If the TO/GA switches are pressed to initiate an automatic go-around, all three autopilots remain in use until another mode selection is made. When this occurs the two autopilots that engaged automatically at 1,500ft are disengaged, as is the rudder channel. This is particularly significant if you happen to be flying on less than all four engines (highly unlikely unless you are in the simulator), as you will have to apply rudder.

Autothrottle system

The autothrottle has two main modes, selected by two switches on the autothrottle section of the MCP: THR (Thrust) and SPD (Speed). In THR mode it sets a fixed thrust; in SPD mode it varies the thrust to maintain airspeed.

In THR REF (Thrust Reference) mode it selects maximum (reference) thrust, as calculated by the FMC and selected on the CDU. The limiting thrust is displayed on the upper EICAS above the NI readout (GE) or EPR readout (P&W and RR).

In THR (thrust) mode it applies thrust to maintain the rate of climb or descent required by the pitch mode.

In SPD (Speed) mode it adjusts thrust to maintain the airspeed selected either automatically by the FMC, by pilot input to the FMC, or directly by the pilot in the speed select window of the MCP.

When HOLD is annunciated the autothrottle is declutched and power can be set manually.

When IDLE is annunciated the thrust levers are moved to the idle position.

There is one mode that annunciates in both pitch and roll modes and also engages the autothrottle, namely TO/GA (Take-Off/Go-Around). As its name suggests it is used for take-off and also in the event of a go-around. It is activated by pressing either of two TO/GA switches, which are attached to the front of thrust levers Nos 2 and 3.

Pressing the TO/GA switches for take-off engages the autothrottle, which advances the thrust levers to set the desired power as shown by the FMC. It also selects 8° of pitch on the flight director pitch bar and annunciates TO/GA in both pitch and roll on the PFDs.

After lift-off the flight director pitch bar commands a pitch to maintain a speed of V2 plus 10kt, which is the initial climb speed. If LNAV and VNAV are selected the pitch and roll annunciations change automatically after take-off, the roll mode at 50ft and the pitch mode at 400ft. The TO/GA mode stays armed until the thrust reduces from take-off power to climb power.

TO/GA mode is rearmed when the first flap selection is made, or the ILS Glideslope is captured. If the TO/GA switches are pressed for a go-around, the autothrottle engages and sets a thrust that will achieve a rate of climb of 2,000ft/min. The pitch bar commands the selected airspeed and TO/GA is again annunciated in both pitch and roll. This time it stays annunciated until other modes are selected. If the autopilots are engaged they will stay engaged and fly the go-around. If the TO/GA switches are pressed a second time, full go-around thrust is selected (which could be quite exciting if the aircraft is relatively light, at the end of a long flight).

(If you have understood all, or even any, of the above you have done well. It is probably time now to put down the book and have a drink!)

Fuel

The aircraft's fuel system is made up of tanks to hold the fuel, electric pumps to pump it under pressure to the engines, and a system of pipes (known as the manifold) for fuel to flow from the tanks to the engines. Four cross-feed valves in the manifold allow fuel flow from any tank to any engine.

The fuel is contained in up to eight different tanks. There are four main tanks in the wings, which are always used, two reserve tanks in

the wingtips, a large centre wing tank (CWT) situated in the centre wing section in the fuselage and additionally there is the option of a tank in the horizontal stabiliser called the Horizontal Stabiliser Tank (HST). Wing tanks 2 and 3 are in the inboard wing sections and are considerably larger than main tanks 1 and 4, which are in the outboard wing sections. Reserve tanks 2 and 3 are actually in the wingtips; fuel is kept in them until the fuel load (and hence aircraft weight) has dropped to predetermined levels, and then transferred automatically to its associated main tank. The fuel is kept in these reserve tanks to keep weight outboard in the wings to reduce wing bending in flight.

There is a surge tank in each wingtip, and one in the right horizontal stabiliser, each having a NACA duct ram air inlet (also called a NACA scoop or a NACA inlet, originally developed

BELOW Fuel system schematic.

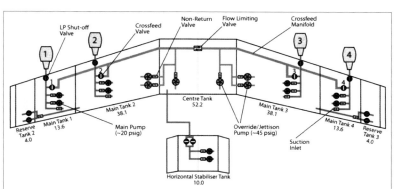

by the US *National Advisory Committee for Aeronautics*, hence the acronym). The vent surge tank has two functions: it acts as a vent for the main fuel tanks, and there are vent lines from each main tank to the vent surge tank. It also acts as an overflow – and allows for over-fuelling or expansion of the fuel in the main tanks (occasionally fuel can be seen leaking through the NACA duct). Any fuel that does get into the vent surge tank drains back in to main tanks 2 and 3. The NACA duct provides positive air pressure to the tanks to allow air to replace fuel as it is used, maintaining a positive pressure in the fuel tanks.

To get the fuel from the tanks to the engines there are two electrically operated fuel pumps in each main tank. Main tanks 2 and 3 have two additional pumps, which operate at a higher

output pressure, which are known as override/jettison pumps. The CWT also has two override/jettison pumps that automatically override the pumps in the main tanks to ensure that the CWT fuel is used before the wing fuel. The HST has two transfer pumps. In the event of fuel pump failure it is possible for the engines to draw fuel from their respective tanks via suction feed lines that bypass the pumps.

The fuel pumps and cross-feed valves are controlled by switches on the fuel management panel on the overhead panel. There are also automatic functions that are controlled by Fuel System Management Cards (FSMCs).

Prior to engine start all the pumps in tanks containing fuel are switched on and the cross-feed valves are switched open. If there is fuel in the CWT this will be supplied to all the engines via the manifold. If the CWT is empty, the override/jettison pumps in tanks 2 and 3 will supply fuel to all the engines. The pumps that are switched on but not operating will operate if a pressure drop is sensed (ie, a pump fails).

When the flaps are extended prior to take-off the FSMCs close the No 2 and 3 cross-feed valves, and CWT fuel is fed to engines 1 and 4. Engines 2 and 3 receive their fuel from their respective tanks, fed by their main pumps. If the CWT is empty engines 1 and 2 are supplied from main tank 2, No 1 by the override/jettison

FUEL TANK CAPACITIES

Fuel tank capacities in US gallons, litres and weight in kilogrammes at a specific gravity (SG) of 0.8 (1 litre of fuel weighs 0.8kg).

Main 1 and 4	8,964 US gal	33,929 L (13,572kg ea)	total	27,144kg
Main 2 and 3	25,092 US gal	94,973 L (37,989kg ea)	total	75,978kg
Reserve 2 and 3	2,644 US gal	10,008 L (4,003kg ea)	total	8,006kg
Total wing fuel	36,700 US gal	138,910 L	total	111,128kg
Centre	17,164 US gal	64,966 L	total	51,973kg
Stabiliser	3,300 US gal	12,490 L	total	9,992kg
Total fuel	57,285 US gal	216,825 L	total	173,460kg

The -400ER version has the option of one or two additional 3,210 US gal fuel tanks in the forward lower cargo hold.

One tank: Total fuel	60,495 US gal	228,990 L	183,192kg
Two tanks: Total fuel	63,705 US gal	241,140 L	192,912kg

pumps, No 2 by the main pumps. Engines 3 and 4 are supplied from main tank 3, No 3 by the override/jettison pumps and No 4 by the main pumps.

When the flaps are retracted after take-off the FSMCs open the No 2 and 3 cross-feed valves. The CWT will now supply all the engines. If the CWT is empty, fuel from tanks 2 and 3 is supplied by their override/jettison pumps to all the engines.

When the fuel in the CWT reaches 36,470kg (which equates to a total fuel load of around 147.6 tons) the FSMCs activate the pumps to transfer fuel from the HST to the CWT.

When the CWT is nearly empty* (the FSMCs detect low-pressure output from the CWT override/jettison pumps) the tank 2 and 3 override/jettison pumps are activated and all the engines are now supplied from main tanks 2 and 3.

(*It is not possible to completely empty the CWT using the override/jettison pumps; around 1,400kg will remain. However, this is transferred to main tanks 2 and 3 by a scavenge system, consisting of four jet pumps.)

When the contents of either main tank 2 or 3 is around 18,140kg (which equates to a total fuel load of around 63.3 tons), the FSMCs open the reserve transfer valves to move fuel from the reserve tanks to their respective main tanks.

When the contents of all the main tanks are equal, which occurs at around 13.5 tons each (a total fuel load of around 54 tons), the EICAS generates a FUEL TANK/ENG message. The override/jettison pumps in tanks 2 and 3 are switched off, cross-feed valves 1 and 4 are closed and each tank now feeds its respective engine. This is known as tank to engine fuel feed and is one of the few operations not done automatically.

If there is a need to reduce the aircraft's weight, fuel can be jettisoned in flight using the fuel jettison system, which utilises the override/jettison pumps and nozzles in each wingtip. As these pumps are only fitted in the CWT and main tanks 2 and 3, fuel from main tanks 1 and 4 and the reserve tanks is automatically transferred into 2 and 3. The reserve tanks transfer when the contents of either main tank 2 or 3 reach 18,140kg, tanks 1 and 4 when main tanks 2 or 3 reach 9,072kg.

Fuel jettison is controlled by a fuel jettison selector switch, a fuel to remain selector and two nozzle valve switches in the overhead panel. Jettisoning is initiated by choosing the jettison selector to A or B (there are two systems), setting the fuel to remain, which is displayed on the EICAS, and opening the nozzle valves. When the fuel quantity reaches the selected fuel level, the system deactivates all the pumps and fuel jettisoning is terminated.

There is a fuelling panel in each wing between the engines, allowing refuelling from either side or both sides simultaneously. There is one fuelling control panel, which is in the left wing. The fuel tanks all contain fuelling valves which are controlled from the fuelling panel.

Fuel distribution will depend on the fuel load, with the wings being filled before fuel is loaded in the CWT, and the CWT filled before fuel is loaded in the HST.

A fuel quantity indicating system (FQIS) uses

a total of 67 sensors in the main tanks, plus a further 10 in the horizontal stabiliser tank, to measure the quantity. The system includes densitometers in all tanks except the reserve tanks. By measuring the volume and the density the system calculates a weight. This is displayed as total quantity on the primary EICAS and also on the secondary EICAS FUEL page, where individual tank quantities are shown.

There is an alternative method of measuring the tank contents that can be used if any tank quantity cannot be displayed. The aircraft is fitted with fuel measuring sticks in each tank: three in each wing tank, one in the CWT and four in the HST. They consist of a float, a measuring stick and stick housing.

Fuel temperature is measured in main tank 1, and displayed on primary EICAS. During fuel jettison this is replaced by an indication of the selected fuel to remain.

BELOW Hydraulic system schematic.

Hydraulics

The 747 has four independent hydraulic systems. The systems are very similar and are the source of power for most of the aircraft's systems. They are designed such that it can still be flown and landed with just one system working.

All the systems are pressurised to 3,000psi, each by a pump driven by its respective engine, known as the Engine Driven Pump (EDP). The EDPs are located on the engines and are controlled by their respective engine's accessory drive. Each system has a secondary pump, known as a Demand Pump, so that in the event of an engine failure, that engine's hydraulic system is still available. Systems No 1 and 4 have air driven pumps whilst systems No 2 and 3 have electric pumps (this is a change from the earlier 'Classic' 747s, where all the

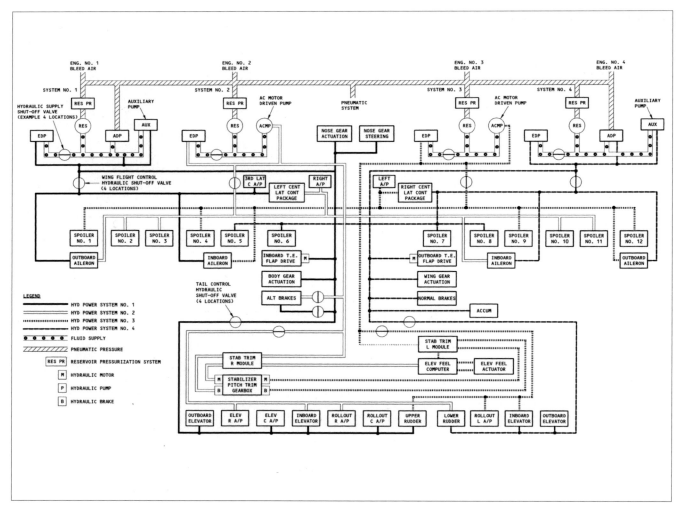

Demand Pumps were air-driven, although this was an option with the -400). The Demand Pumps, along with most other components of the hydraulic systems, are mounted in their respective engine struts.

There is also an electrically powered auxiliary pump for system No 4, which is only used on the ground. With system No 4 powered, the brakes are available during start and towing operations. Some aircraft have an additional auxiliary pump on system 1, which means that the nose and body gear steering can also be powered during towing.

The hydraulic fluid is stored in reservoirs, one for each system, mounted in their respective engine struts. From the reservoirs the fluid flow is split between the pumps, with flow to the EDP passing through a shut-off valve. After passing the pumps the fluid travels through filters, passing pressure switches on the way. After the filters the fluid passes a pressure transmitter before being supplied to the system. There is also a relief valve which allows excess fluid to return to the reservoir.

Switches for all the systems are located on the left-hand side of the P5 overhead panel in the flight deck. The EDPs have push-button switches which are normally in the ON position. The Demand Pumps have three-position rotary switches, OFF, AUTO and ON. OFF and ON are self-explanatory; if switched ON the pump runs continuously, if switched OFF it is switched off! In the AUTO position the pump will be switched on if its EDP output pressure is below 1,400psi for 0.3sec and it will run until it has maintained a pressure above 1,600psi for 14sec. It will also run if its associated engine's fuel metering valve is closed (ie the engine is shut down). Additionally the No 1 and 4 Demand Pumps also run when the trailing edge flaps are in transit or out of the UP position when airborne.

The system 4 Demand Pump switch (and additionally on some aircraft system 1) also has an AUX position, which switches on the electrical auxiliary pump.

Before start-up the Nos 1, 2 and 3 Demand Pumps switches are set to AUTO. As the engines are not running the Demand Pumps will run to provide system pressure. As the engines start and the EDPs start producing pressure, the Demand Pumps will automatically switch

ABOVE Hydraulic fluid refill panel inside the left body gear wheel well. *(Chris Wood)*

off. If an engine should fail or be shut down, as system pressure falls the Demand Pump will automatically switch on to restore pressure.

Located in the centre of the overhead panel are the engine fire switches. One of the functions of these switches is to shut off the hydraulic fluid supply to the EDPs.

The hydraulic systems are controlled by four hydraulic interface modules (HYDIMs), each system having its own circuit card located in the MEC. The HYDIMs receive signals from the switches as well as pressure and temperature information from the pumps. They process the information and send control messages to the pumps and valves. They also feed information to the annunciator lights in the switches, the EICAS system and the CMC.

Also in the MEC is a hydraulic quantity indicating module (HYQUIM), which receives information from probes and transmitters in the hydraulic fluid reservoirs and sends signals to the EICAS, the CMC and the SYS FAULT lights in the pilots' overhead panel.

The SYS FAULT light illuminates if the system pressure is low, the fluid quantity is low or the fluid temperature is high.

The load is split between the four systems, with all the systems having some input into the flight controls. Whilst the aircraft cannot be flown without any hydraulics, as long as any one system is operating some control can be maintained and the aircraft landed safely. Systems 1 and 4 primarily power the landing gear, brakes, steering and the trailing edge flaps. System 1 powers the inboard gear and flaps (the number '1' is like the letter 'I' – for

ABOVE Hydraulic fluid refill panel inside the left body gear wheel well. *(Chris Wood)*

Inboard – easy to remember), system 4 the outboard gear and flaps.

The wheel brakes have three sources of hydraulic power: system 4 is the normal source with system 1 as a back-up and system 2 as an emergency back-up.

System 1	System 2	System 3	System 4
Centre autopilot	Right autopilot	Left autopilot	
Nose and body gear retraction & extension			Wing gear retraction & extension
Nose and body gear steering			
Alternate brakes	Alternate brakes		Normal brakes
Inboard trailing edge flaps			Outboard trailing edge flaps
Left inboard elevator	Left inboard elevator	Right inboard elevator	Right inboard elevator
Left outboard elevator			Right outboard elevator
	Elevator feel	Elevator feel	
	Stabiliser trim	Stabiliser trim	
Upper rudder	Lower rudder	Upper rudder	Lower rudder
Left inboard aileron	Right inboard aileron	Left inboard aileron	Right inboard aileron
Left outboard aileron	Left outboard aileron	Right outboard aileron	Right outboard aileron
	Spoilers	Spoilers	Spoilers
P&W engine only:			
No 1 engine thrust reverser	No 2 engine thrust reverser	No 3 engine thrust reverser	No 4 engine thrust reverser

The hydraulic servicing station is in the left body gear well, and all the systems can be topped up from this one point. There is a pressure connection for a ground cart, which is the normal method for replenishing the systems, a hand pump with a suction line, a selector valve and a quantity indicator. Selecting the valve to a particular system activates its quantity gauge, as well as routing the fluid to that system. The reservoirs are located in the engine struts, and there are various filters throughout the systems.

Electrics

The aircraft generates its own electrical power using four 90 KVA Integrated Drive Generators (IDGs), one attached to each engine driven by the N2 accessory drives, to generate 115V 400Hz ac power. Each IDG contains a generator, a constant-speed drive system, a lubrication system and control and protection circuitry. They are automatically synchronised so that they can be run in parallel.

The constant-speed drive in the IDG corrects for changes in input speed to keep the generator running at a constant 12,000rpm to produce 400Hz. The IDG oil is partially cooled by a fuel/oil heat exchanger, supplemented by air/oil heat exchangers. If an IDG develops a fault its drive can be disconnected, but only whilst the engine is running. In this event the generator output will be lost, and the IDG can only be reconnected on the ground.

To provide electrical power on the ground there are two 90 KVA generators fitted to the Auxiliary Power Unit (APU), as well as two sockets where a 90 KVA ground electrical supply can be plugged in. The APU runs at a constant speed so its generators do not need constant-speed drives, the APU generators running at 8,000rpm to produce 400Hz. The aircraft can be powered by the APU generators, the ground power or it can be split between the two.

This ac power is distributed to the various electrical services via a system of busbars, each electrical item receiving its power from a busbar (they are normally referred to as bus or busses – pronounced *buzz*). Each generator is connected to its own busbar via a Generator Control Breaker (GCB) and all the main busbars are connected via Bus Tie Breakers (BTBs) to a synchronous bus. This means that if a generator fails, an IDG has to be disconnected or an engine fails, no electrical services will be lost because the associated busbar will still be powered from the synchronous busbar.

The synchronous bus is split into two halves by the Split System Breaker (SSB). Neither the APU generators nor the ground electrical supplies are synchronised so they cannot be run in parallel. On the ground the SSB splits the system in two, so that each APU generator or ground supply is powering half of the system. If there is only one supply, or there are two and one fails, the SSB automatically closes so that the whole system is powered.

There are further ac busbars; 2 transfers busses, 2 standby busses, 4 galley busses, 4 utility busses and a ground servicing bus. There is also a ground handling bus, which is only powered on the ground.

The pilot's flight instruments are powered from a pair of transfer busses. The Captain's transfer bus normally gets its power from ac

bus No 3, the First Officer's from ac bus No 2. If their normal power sources fail, these busses automatically transfer to ac bus No 1.

The four galley busses provide power to the aircraft's main galleys, and the utility busses power various aircraft systems, predominantly items that draw a heavy electrical load such as fuel pumps. Each one is powered by its respective main ac bus. There is a pair of utility bus switches in the flight deck overhead panel; the left switch controls utility busses No 1 and 2 and galley busses No 1 and 2, the right switch controls utility busses No 3 and No 4 and galley busses No 3 and No 4.

The ground servicing bus is powered in flight or on the ground by ac bus No 1. On the ground it can also be powered directly from the power source for the ground handling bus via a switch by the L2 door.

The ground handling bus is only available on the ground and is powered directly from either the No 1 APU generator or the No 1 ground electrical supply. If both are available the ground supply has priority. The power doesn't actually

BELOW Electrical system schematic.

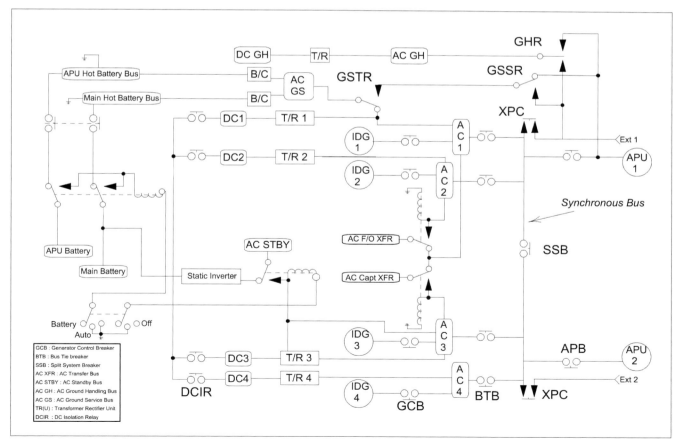

via their battery charger, which keeps the battery fully charged and also acts as a TRU for it.

There is a standby power system that provides ac power in the highly unlikely event of a total loss of main electrical power. It automatically provides power to some flight instruments, navigation and communication equipment for at least 30mins. It uses dc power from the two batteries and converts it to ac using two static inverters, which supply two standby busbars. The main battery supplies power to the main standby inverter and the main standby busbar; the APU battery supplies the APU standby inverter which supplies the APU standby busbar.

The electrical system is controlled through four generator control units (GCUs), one per generator, two auxiliary generator control units (AGCUs), one per APU generator and two bus control units (BCUs). These provide automatic power transfers, load management, with automatic load shedding if required, and system protection.

The GCUs regulate the voltage, direct the loading and provide automatic control and protection for their respective channels. They also supply the indications for EICAS and the lights in the electrical system switches in the overhead panel. The BCUs control and protect the synchronous busbar, and in conjunction with the GCUs through ARINC 429 data links they provide load management, SSB control, power transfers and load splitting for autolands.

The Freighter aircraft have a main-deck cargo handling bus. It is only powered on the ground, by either the No 2 APU generator or the No 2 ground electrical supply. If both are available priority is given to the ground electrical supply. The power has to be available (AVAIL light illuminated), but not selected ON, for the bus to be powered.

need to be selected on for the ground handling bus to be powered.

To convert ac power to dc power there are four 75-ampere Transformer Rectifier Units (TRUs) supplying four dc busbars, each TRU being powered from its respective ac busbar. The TRUs convert 115V ac to 28V dc. There are also two 20-ampere TRUs, powered by the ac ground handling busbars that supply two dc ground handling busbars. The TRUs are in the MEC.

Like the ac busbars, the dc busbars are interconnected, through dc Isolation Relays (DCIRs) to a dc tie busbar. The DCIRs are controlled by the BTBs, so if a BTB is selected open its main ac busbar is isolated from the synchronous busbar and its dc busbar is also isolated from the dc tie busbar. If an ac busbar is automatically isolated due to a fault, the BTB opens but the DCIR does not.

There are two 28V batteries, a main aircraft one mounted on the right-hand side of the MEC and an APU one in the Aft Equipment Centre. They are nickel cadmium batteries, each rated at 40-ampere hours. They both have their own battery chargers and they both have a pair of associated busbars, a battery busbar and a hot battery busbar.

The battery busbars are normally powered by dc busbar 3, but if this is unpowered they can be operated from the hot battery busbars, provided the battery master switch is on.

The hot battery busbars are connected directly to their respective batteries. They are normally powered by the ground service busbar

Pneumatics

The pneumatic system supplies pressure and temperature-regulated air for:

- Engine starting
- Air conditioning and pressurisation
- Operation of the leading edge flaps

- Operation of the engine thrust reversers (GE and RR engines)
- Hydraulic Demand Pumps (Nos 1 and 4 hydraulic systems)
- Wing and engine anti-icing
- Pressurisation of the hydraulic reservoirs
- Pressurisation of the water tanks
- Heating of the aft cargo compartment
- Cargo smoke detection

The normal air source is engine bleed air. For engine starting APU bleed air can be used, or air from an external ground air cart.

Air is bled for the compressor section of the engine. For the GE engine, Intermediate Pressure (IP) 8th-stage air is normally used, through a check valve, supplemented by High Pressure (HP) 14th-stage air via the High Pressure Shut-Off Valve (HPSOV) at low power settings. The air passes through a Pressure Regulating Valve (PRV) and a pre-cooler, which uses cool air bled from the fan section of the engine through the Fan Air Modulating Valve (FAMV) to regulate the bleed air temperature. Upstream of the pre-cooler are the engine thrust reverser, the engine start valve, the nacelle thermal anti-ice valve and the Pressure Regulating and Shut-Off Valve (PRSOV). The section of manifold before the PRSOV is the engine manifold, beyond the PRSOV the air enters the main pneumatic manifold.

The pneumatic manifold runs along the wing leading edges and through the air conditioning pack bays in the lower fuselage, where it is referred to as the crossover manifold. There are two isolation valves in the crossover manifold, one in each air conditioning pack bay, which enable isolation of sections of manifold. The APU bleed air is introduced to the crossover manifold through an APU bleed air valve situated at the rear of the left body gear wheel well. Ground conditioned air is introduced to the crossover manifold through two ground connectors in the left air conditioning pack bay.

Air is taken from the manifold at various places to power the air-driven systems. Manifold pressure is displayed at the bottom of the upper EICAS (as duct pressure) and on the lower EICAS on the Environmental Control System (ECS) synoptic.

Engine bleed air switches in the flight

BELOW The leading edge flaps are normally operated by the pneumatic system. They are all deployed here on British Airways' G-BNLS as it nears runway 27L at Heathrow in September 2009. *(Chris Wood)*

deck overhead panel enable the PRSOV, the PRV controller and the HPSOV controller. These valves are electrically controlled but pneumatically powered, ie they may be switched on, but if there is no air pressure they will be closed.

For engine starting, the PRSOV opens to allow reverse flow to the engine starter motor and the PRV is closed to prevent reverse flow into the engine compressor section. (The bleed air switch must be switched on, to provide power to the PRSOV and the PRV.)

Air conditioning

The air conditioning system provides a mixture of hot engine bleed air and recirculated cabin air at a controlled temperature into the passenger cabin and flight deck. Hot air from the pneumatic manifold is fed into three air conditioning packs via flow control and shut-off valves. Here it is cooled before passing into a plenum where the output from the three packs is mixed prior to being fed into ducting that circulates it around the aeroplane. For

distribution the aeroplane is divided up into seven zones, five on the main deck and two on the upper deck.

The air conditioning packs consist of a dual heat exchanger and an Air Cycle Machine (ACM), which consists of a compressor and a turbine. From the valve the air flows through the first stage of the heat exchanger, which cools it. It is then compressed by the compressor, which also warms it; this heat is removed by the second stage of the heat exchanger. It then moves through the turbine, which cools it further, and it is then passed through a water separator before being supplied to the plenum.

Cooling air for the heat exchanger is provided by ram air in flight, which enters through a ram air inlet and exits through a ram air exhaust. On the ground a fan in the ACM draws air into the heat exchanger. Pack outlet temperature is managed by a pack temperature controller, which directs the position of the ram air inlet, the ram air exhaust and flow through the ACM turbine. It regulates the latter by controlling flow through a turbine bypass by an associated turbine bypass valve.

The packs are fitted in the underside of the aircraft, Nos 1 and 2 on the left-hand side and No 3 on the right and can be accessed through doors.

The packs are controlled by three switches in the flight-deck overhead panel, marked OFF, NORM, A and B. There are two pack temperature controllers, A and B, and each has three channels, one for each pack. Only one will be in operation, but they are normally scheduled to change over on touchdown (so for one flight A is in control, the next it's B, then A etc). If one channel fails the other controller can be selected. The packs have two flow settings, normal and high. In normal operation they automatically operate in high flow on the ground, in the climb and in descent. However, pack high flow can also be selected ON by the high flow switch on the overhead panel.

Temperature control for the cabin is achieved by mixing the cool pack outlet air with hot air bled directly from the pneumatic manifold, known as trim air. It is controlled by two temperature selectors in the flight deck, one for the flight deck and one for the passenger cabin. Temperatures for the individual zones in the cabin are controlled at the panel at the R2 door. The zone set to the coldest temperature controls the output temperature from the packs. For the other zones, demanding higher temperatures, trim air is added through seven trim air modulation valves.

For distribution the air is ducted from the plenum into a manifold that supplies it to six sidewall risers, which is where the trim air is added. Air for the flight deck is tapped off the riser for zone E, which is at the rear of the main deck. From the risers it is ducted to overhead distribution ducts connected to diffusers in the cabin sidewall.

There are two pairs of recirculation fans, one pair above the cabin roof and one below the cabin floor, which recirculate cabin air by adding it to the plenum output. Additional recirculated air can be supplied to the cabin by the gasper system. This recirculates air from above the cabin and supplies it to gasper outlets at the PSUs.

Air leaves the cabin though floor-level grills, and is routed around the cargo bays and aft to the area behind the bulk cargo compartment. Here it is discharged overboard through two outflow valves. Air from the galleys and lavatories is drawn out by a pair of vent fans, a primary fan and a back-up.

For conditioning on the ground it is possible to connect ground conditioned air, which can be from an air cart or a fixed ground installation. There are two connectors downstream of the water separators for packs 1 and 3. They can be accessed through the associated pack access panels in the underside of the aircraft. Ground conditioned air is particularly useful when operating in a hot climate with an unserviceable APU.

Equipment cooling

The electronic equipment in the flight deck and the MEC generates a lot of heat. To cool it air is drawn from the cabin through an equipment supply valve using a pair of fans, a supply fan and an exhaust fan. This air is then exhausted either into the forward cargo compartment for heating, through an inboard exhaust valve, is dumped overboard through a ground exhaust valve, or it is re-circulated.

There is a three-position equipment cooling selector on the overhead panel, with STBY,

LEFT Passenger cabin upper-deck air conditioning riser pipe work, normally concealed behind the cabin wall panelling. *(Jonathan Falconer)*

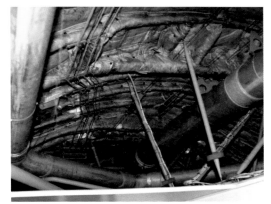

LEFT Overhead air conditioning distribution ducting above the main passenger cabin false ceiling. *(Chris Wood)*

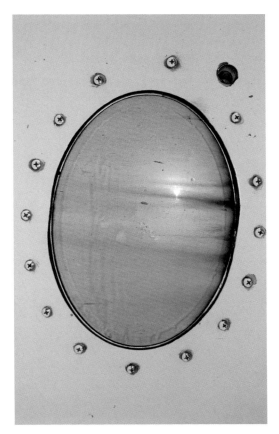

LEFT Electrical equipment ground exhaust valve. *(Chris Wood)*

NORM and OVRD settings. In NORM mode both fans operate; on the ground with no engines running with a warm temperature, the air exhausts overboard. At colder temperatures, and once an engine has been started on both sides, the ground exhaust valve closes and the air exhausts into the forward cargo compartment. If the system detects a fault with one of the fans, the inboard exhaust valve closes and the system reconfigures to become recirculating. With STBY selected the automatic functions are bypassed and the system configured manually for flight; the ground exhaust valve closes and the inboard exhaust valve opens.

If OVRD is selected both fans are switched off, the supply, inboard exhaust and ground exhaust valves close and a smoke/override valve opens. Cooling air is provided by differential pressure, so this should not be used on the ground.

Cargo heating

Both the forward and aft cargo compartments are heated. The forward compartment is heated by equipment-cooling exhaust air from the flight deck and the MEC, supplemented by two electric heaters in the exhaust ducts.

The aft and bulk compartments are heated by air from the pneumatic manifold drawn from the crossover manifold, controlled by a switch in the overhead panel. The temperature is maintained at either 45°F or 65°F, as selected in the bulk cargo compartment by a switch forward of the door.

Pressurisation

The pressurisation system manages the discharge of conditioned cabin air through two outflow valves in the rear of the fuselage. It is controlled automatically by a pair of cabin pressure controllers in the MEC. In normal operation only one controller is directing, the other acting as a back-up in case of failure and the controllers alternate between flights.

If the automatic function fails the outflow valves can be regulated manually using switches on the flight deck overhead panel.

The system uses the inputted Flight Management Computer (FMC) flight plan to calculate a pressurisation schedule.

Cabin altitude controls are on the overhead panel. There is a three-position cabin altitude auto selector, marked AUTO, A and B, two outflow valve manual switches for selecting the valves to manual, an outflow valve manual control switch which moves them and two outflow valve position indicators.

The landing altitude is normally set automatically by the FMC, but it can also be selected manually using the landing altitude selector.

To prevent overpressurising the aircraft, two mechanical pressure relief doors are fitted to the lower fuselage on the left-hand side. These open if the pressure reaches 9.4psi. If either one opens, pack 2 is automatically switched off.

If cabin pressure is higher than the ambient pressure, two pairs of negative-pressure relief doors in the forward and aft cargo compartment doors open. These also open as the doors are unlatched, to relieve any remaining differential pressure.

Oxygen

The aircraft has two independent oxygen systems, one for the flight crew and one for the passengers. The oxygen is stored in cylinders mounted on the right-hand sidewall of the forward cargo compartment, each cylinder

RIGHT This pressurised cylinder inside the cargo hold contains 115cu ft of oxygen. *(Chris Wood)*

has a pressure gauge and a valve, which is normally open. System pressures are shown on the lower EICAS status page. The systems can be topped up either by changing cylinders or through a service panel mounted in front of the cargo door inside the forward cargo compartment. Each cylinder has a frangible disc; if pressure becomes excessive the disc will be ruptured and the oxygen discharged overboard. There is a discharge port just aft of the forward cargo door; this is fitted with a green indicator disc which will be blown out by the excess pressure.

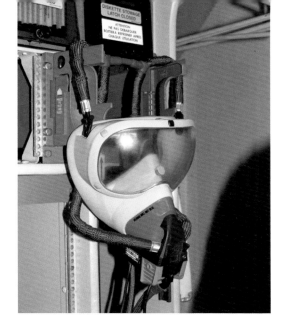

LEFT Crew oxygen mask. *(Chris Wood)*

Flight crew oxygen

The flight crew oxygen system is a demand system, it only flows when demanded, and it normally consists of two cylinders. From the valve the oxygen flows through a pressure reducer and a pressure regulator to a low-pressure distribution manifold which is connected to quick-donning full face masks, one at each of the four pilot stations. The masks are fitted with diluter demand regulators, the diluter function enables breathing of an air/oxygen mixture, the demand function means that oxygen only flows when demanded, ie, the wearer inhales. A NORMAL/100% switch on the regulator allows selection of an air/oxygen mix, with a ratio determined by the cabin altitude, or 100% oxygen. A rotary Emergency/Test selector when pressed tests for a positive supply pressure; when rotated to Emergency it supplies 100% oxygen under pressure.

Oxygen flow to the mask is activated when the mask is removed from its stowage box, and flow is demonstrated by a yellow flow indicator. Each mask is fitted with a microphone; this is activated when the left hand door of the mask stowage box is opened.

The system can be shut off by closing the left-hand door of the mask stowage box and pressing the RESET/TEST slider, fitted on the left-hand side. This also switches off the mask microphone.

Passenger oxygen

The passenger oxygen system is a continuous flow system; once activated oxygen flows until the system is switched off. The number of passenger oxygen cylinders fitted varies

LEFT Flight-deck oxygen mask panel. *(Chris Wood)*

depending on customer choice. From the cylinder valves the oxygen flows through a pressure reducer to three continuous flow control units, mounted behind the oxygen cylinders in the forward cargo compartment. When selected on, oxygen flows from here to a low-pressure manifold.

The low-pressure manifold distributes the oxygen to the passengers' masks in the Passenger Service Units (PSUs), to the cabin crew stations, to the crew rest areas and to lavatory service units. Flow to individual masks is only initiated when the mask is pulled sharply downwards, this opens a valve in the PSU.

The passenger oxygen system is switched on automatically when the cabin altitude exceeds approximately 14,000ft (quoted as 13,250ft to 14,500ft) or it can be turned on by a switch on the overhead panel in the flight deck. The automatic activation is triggered by a pressure sensor in the flow control units. When they open there is an initial pressure surge which opens all the panels housing oxygen masks, and purges air from the manifold.

System pressure is regulated depending on cabin altitude; it is approximately 2psi at 14,000ft and 43psi at 40,000ft.

Once the cabin altitude is below 12,000ft the system can be reset by changing the flight-deck switch to the spring-loaded RESET position. This closes the flow control units.

Therapeutic oxygen

The aircraft can be fitted with an additional therapeutic oxygen system. This is part of the passenger oxygen system and allows use of the passenger oxygen without having to activate the whole system, using masks plugged in to therapeutic oxygen outlets on the PSUs. It is activated by a Therapeutic Oxygen switch in the flight deck but requires at least one therapeutic oxygen mask to be connected before activation. Therapeutic oxygen is also activated automatically in the event of the cabin altitude reaching 14,000ft.

It can be turned off with the flight deck switch once all the masks are unplugged.

Water

The aircraft has a potable water system that supplies water to the galleys, lavatories and a humidifier (if fitted). The water is stored in three 110US gal fibreglass tanks mounted on the front spar of the wing centre section at the rear of the forward cargo compartment. The system is pressurised from the pneumatic manifold, or from a separate electrical compressor if the pneumatic pressure is low.

There is a water service panel in the underside of the aircraft, forward of the main wheel well, which is used to refill and drain the system, and a water drain panel behind the main wheel well, also used for draining.

Water system quantity is shown at the water service panel and in the cabin at door R2.

Waste water from the galleys and lavatory sinks is drained overboard through three heated drain masts on the underside of the aircraft.

Waste

There are lavatories with toilets located around the aircraft, on both the main and upper decks. The toilets use a vacuum system to extract waste and send it to the waste storage tanks. The vacuum system uses aircraft cabin differential pressure in the air above 16,000ft, and a vacuum blower system below 16,000ft (below 12,000ft on descent) and on the ground.

Toilet flushing is controlled by a flush control unit. Pressing the flush lever signals the flush control unit to activate the vacuum blower for 15sec. One second after the lever is operated, the flush control unit opens a rinse water valve, which stays open for 1sec. Two seconds after moving the lever the flush valve opens, allowing the contents to be sucked into the pipe taking them to the waste tank. The flush valve stays open for 4sec. If the flush valve fails open it can be closed by hand using a manual handle.

There are four waste tanks mounted on the bulk cargo compartment sidewalls. These are made of fibreglass with a stainless steel liner, and there is an 85US gal tank and a 65US gal tank on each side. Each tank has two level sensors; when they both sense that the tank is full, flushing is inhibited at the associated toilets and LAV INOP lights are illuminated at the panel at the R2 door.

The toilets are serviced at the waste service panel, which is situated under the fuselage near the rear of the aircraft.

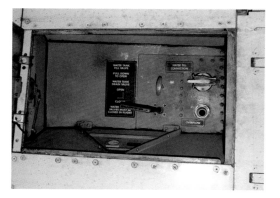

RIGHT Potable water service panel. *(Chris Wood)*

FAR RIGHT Potable water tanks inside the cargo hold. *(Chris Wood)*

Protection systems

Ice and rain protection

Ice and rain protection is achieved using Thermal Anti-Icing (TAI) and electrical systems. The TAI system uses engine bleed air to anti-ice the engine nacelles and the wing leading edges. Electrical heating is provided to flight-deck windshields, the four pitot-static probes, two Total Air Temperature (TAT) probes, two angle-of-attack sensors, water and waste lines and three drain masts.

Thermal anti-ice – engine

Bleed air from each engine is fed to the leading edge of the engine's nacelle, through a Nacelle Anti-Ice (NAI) valve. Switches on the overhead panel, one per engine, control the valves. The hot bleed air is fed into a spray duct inside the nacelle leading edge and is vented overboard. When the valves are open green NAI indications display on the upper EICAS.

When an engine's nacelle anti-ice is switched on, that engine's igniters are also switched on and operate continuously.

Thermal anti-ice – wing

Bleed air from the pneumatic manifold is used to anti-ice the wing leading edges. Wing anti-ice is activated by a single switch in the flight deck overhead panel, which controls a pair of Wing Anti-Ice (WAI) valves, one in each wing. The hot air is fed to spray tubes that are perforated to allow the hot air to escape inside the leading edge. The air vents through six outlets on the underside of each wing.

When the valves are open green WAI indications display on the upper EICAS, one per valve.

Wing anti-ice is inhibited on the ground, and when the leading edge flaps are extended.

Some aircraft have automatic ice detectors fitted. They also have an automated function for their nacelle and wing anti-ice that automatically switches them on if ice is detected.

Electric heating

The four pitot-static probes are electrically heated. They are automatically powered whenever any engine is running, at a reduced heat when on the ground. This switches to full

heat when airborne. The two angle-of-attack sensors are heated when any engine is running, the two TAT probes are heated when the aircraft is airborne.

All the water and waste lines that are at risk of freezing are heated electrically using a combination of inline heaters, heater tapes and integrally heated hoses. Water from the galleys and lavatories drains through three electrically heated drain masts. These operations are all automatic and are controlled by air/ground logic.

ABOVE Drain mast. *(Chris Wood)*

BELOW From the top: angle of attack sensor; two pitot static probes; TAT probe. *(Chris Wood)*

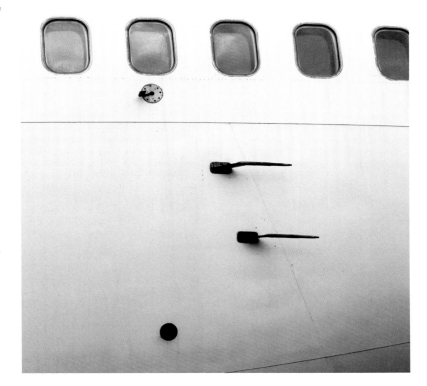

Windshields

(See Chapter 2 – Anatomy for full details.)

Fire protection

The 747 has fire detection and extinguishing systems for the engines, APU, cargo holds and lavatories. It also has fire detection systems in the wheel wells and smoke detection in the crew rest area. All fire warnings generate a red EICAS message, illuminate the red master warning switches on the glare shield panel and activate the fire warning bell. This rings on for two seconds and off for three and can be cancelled by pushing either master warning switch (or putting out the fire!).

Engine fire detection and protection

Each engine nacelle is fitted with a dual loop overheat detector and a dual loop fire detector. Both systems normally require detection in both loops to trigger a warning. However, the systems self-test, so if they detect a failure in one loop they reconfigure for single-loop operation.

Detection of an overheat illuminates an EICAS caution message, and indicates a hot air leak from or around the engine.

Detection of an engine fire also illuminates a red light in the associated engine fuel switch, a red light in the associated engine fire switch and electrically unlocks the switch so it can be pulled out.

The engine fire switch is unlocked by an engine fire warning, by moving the engine fuel switch to cut off, or by pressing the manual override lock button behind it.

Pulling the engine fire switch shuts down the engine and:

- Closes the associated engine and spar fuel valves
- Disconnects the associated electrical generator
- Closes the associated engine bleed air valve
- Shuts off the hydraulic fluid to the associated engine-driven hydraulic pump and depressurises it
- Arms the associated engine fire extinguishers

The fire extinguishers are discharged by rotating the engine fire switch to left or right. They discharge through four nozzles around the engine fan and core. If the fire warning is still illuminated after firing the first bottle, rotating the fire switch the other way will discharge the second bottle.

The fire extinguisher bottles are mounted in the wing leading edge and can be accessed by lowering the leading edge flaps. Aircraft with P&W or GE engines have two fire extinguishers per wing, and they can be discharged either both into one engine or one per engine. RR-powered aircraft have two bottles per engine, which only supply their respective engine.

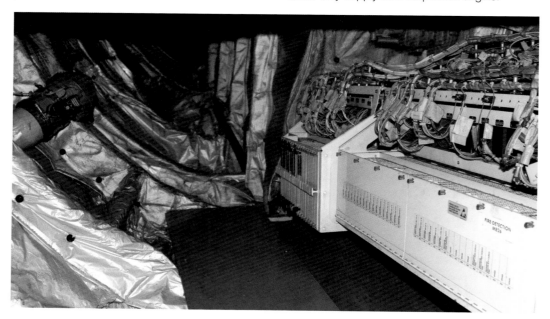

RIGHT Fire detection electrical equipment in the MEC bay.
(Chris Wood)

APU fire detection and protection

The APU compartment also has a dual loop fire detection system. On some aircraft detection in both systems is required to activate a warning once an engine is running. On these aircraft when no engines are running, and at all times on other aircraft, only one loop detecting a fire will activate a warning.

An APU fire warning will also activate a fire warning bell at the APU ground control panel in the right-hand body gear wheel well, illuminate the APU fire switch on the overhead panel in the flight deck and automatically shut down the APU. On the ground it will also automatically discharge the APU fire extinguisher bottle.

The APU fire switch is unlocked in the same way as the engine fire switches. Pulling the APU fire switch shuts down the APU and:

- Closes the APU fuel valve
- Disconnects the APU generators
- Closes the APU bleed air valve
- Arms the APU fire extinguisher bottle

The single APU fire extinguisher bottle is mounted on the left-hand side in front of the APU firewall.

Cargo fire detection and protection

Both the forward, aft and bulk cargo compartments are fitted with a pair of dual loop smoke detection systems; both loops are required to detect smoke to activate a cargo fire warning. A venturi ejector, under the floor at the forward end of the aft cargo compartment, is connected to the pneumatic manifold and draws air across the smoke detectors. The smoke detectors consist primarily of a light, a light trap and a photo cell. The light is permanently on, and shines at the light trap. If smoke is present the light is reflected from smoke particles into the photocell, activating the fire warning.

Cargo fire protection is provided by four fire extinguisher bottles, which are fitted on the right-hand sidewall of the forward cargo compartment. The bottles are discharged using switches in the flight-deck overhead panel. There are two cargo fire arm switches, one for the forward

compartment and one for the aft, and a cargo fire discharge switch. To discharge the bottles, press the required fire arm switch and then the discharge switch. Two bottles discharge immediately, the remaining two start to discharge 30 minutes later, or on landing if sooner.

Wheel well fire detection

The wing and body gear wheel wells have a continuous loop fire detector system with eight sensors, two in each gear wheel. If an overheat is detected it will activate the warnings, but there is no protection system (the normal procedure is to lower the landing gear).

Lavatory fire detection and protection

The lavatories are fitted with smoke detectors. These will illuminate a red light and activate an aural warning when sensing smoke. Fire extinguisher bottles are fitted in the lavatory waste bins. They have two discharge nozzles which are sealed with solder. In the event of a fire, the solder will melt, discharging the extinguisher automatically.

Leading edge overheat

The wing leading edge area contains ducting for hot engine bleed air. To detect leaks there are thermal switches located in the wing leading edges, the engine struts, the air conditioning pack bays and associated ducting. There are nine per wing, two per engine strut, six in the right air conditioning pack bay, four in the left and a further five near the crossover duct (the section in the fuselage). If they detect an

Chapter Five

Flying the Boeing 747-400

Captain Chris Wood

The 747-400 is a highly automated two-crew aircraft, so operating it is a very different experience to the three-crew Classic 747. However, the handling is actually not that different and the 747 is, and always has been, a relatively easy aircraft to fly considering its size. It guarantees a smile after every landing!

OPPOSITE Take-off. The nose gear doors are about to close. *(Shutterstock.com)*

Documentation

Not surprisingly, a machine as large and complex as a 747 comes with a lot of paperwork! For the pilots there are a plethora of manuals, based around the Federal Aviation Authority (FAA) Airplane Flight Manual (AFM). This document is contained in three large A4-size files and details the aircraft's limitations (such as maximum weights, maximum speeds, engine limitations etc) but is mostly made up of performance graphs.

There is a Flight Crew Operations Manual (FCOM) which is split into two parts. Part 1 describes the procedures for operating the aircraft and Part 2 provides technical information on the various aircraft systems. There is also a Flight Crew Training Manual (FCTM) which gives more detailed information on how to actually fly the aircraft.

There is a large ring binder known as the Quick Reference Handbook (QRH) which contains the Normal Checklists (NC) and the Non Normal Checklists (NNC).

The engineers' main document is the Aircraft Maintenance Manual (AMM). This details all the maintenance functions that the engineers have to carry out, from such things as changing light bulbs, washing the aircraft, to changing an engine. Each maintenance action is detailed in a series of procedural steps.

Backing this up is a document known as the Aircraft Illustrated Parts Catalogue (AIPC).

Each aircraft has its own Technical Log Book (referred to as the Tech Log). This document is a history of the aircraft's operational service, with every flight and every maintenance action being recorded in it. It is divided into several sections, but mostly consists of pages recording the aircraft's flights. Over the aircraft's lifespan these add up to thousands of pages so only the most recent ones are kept in the log. Every defect is recorded in the log and each has to be rectified or deferred before the next flight. If the defect is rectified the actions taken are recorded, if it is deferred the details of the defect are transferred to the Deferred Defects section, along with a reference from the Despatch Deviation Procedures Guide (DDPG) or Minimum Equipment List (MEL) for referral.

Other sections of the Tech Log contain a list of the deferred defects, any extra maintenance actions required (usually as a result of deferred defects) and their periodicity.

The DDPG lists all the defects that can be deferred. It is rare for a large aircraft to be completely 100% serviceable but various equipment and system failures are acceptable for flight, especially in an aircraft like the 747 that has four of most things! The DDPG lists all the systems that the aircraft can be operated without, along with procedures for the engineers to disable the systems and procedures for the pilots to operate the aircraft without them. This may include extra limitations or different operating procedures and may also include a time limit on how long the aircraft can be operated with the defect before it must be repaired. For example, if an anti-collision beacon is not working, it is permissible to despatch the aircraft for flight provided that the strobe lights are working. If the No 1 or No 4 Hydraulic Demand Pump is not working, it is permissible to despatch the aircraft provided that gear down take-off performance is used. Why? Well, if the associated engine fails during take-off (a highly unlikely event!) before the landing gear is retracted, that engine's hydraulic system will lose its pressure because its back-up pump is not working. Consequently part of the landing gear will not retract, creating more drag and compromising the aircraft's climb performance. In this situation, by assuming that all the landing gear will not retract when calculating the take-off performance, an extra layer of safety is built in. Engine failures are extremely rare but there is a law (Sod's) that states that things happen when you least expect, or want them to!

As well as these Boeing-supplied manuals there will be a number of the operator's own manuals outlining the Company's rules and procedures for operating the aircraft, as well as other paperwork such as its Certificate of Airworthiness issued by the country's regulatory authority (the European Air Safety Agency in Europe) and its insurance certificate.

In addition there will be files of maps and charts for the pilots to refer to showing departure, arrival and approach procedures. These, and all the paper manuals can, however, be replaced by a device called an Electronic Flight Bag (EFB), which is effectively a laptop for pilots.

Payload and range

There are many figures banded around giving maximum ranges and typical seating capacities for the 747. It is best to delve a little deeper into this area to gain an understanding of what the aircraft can do.

The empty weight of a typical passenger -400, fully prepared to fly but without any fuel or payload onboard is around 180,000kg (396,832lb). This weight will vary between aircraft depending on such things as the engine fit, seating and galley configurations and even the In Flight Entertainment (IFE) system.

The aircraft has several certified maximum weights which vary from version to version but are laid down in the aircraft's AFM. It is not permissible to exceed these. The standard

BELOW Cargo off-loaded from Air France 747-400, F-GITD, at San Francisco on 29 April 2011. *(Chris Wood)*

passenger -400 has a Maximum Take-Off Weight Authorised (MTWA) of 394,625kg (870,000lb). This is a structural limit and takes no account of things such as the length of the runway.

There is a certified Maximum Zero Fuel Weight (MZFW). It will come as no surprise to you to learn that this is the maximum weight of the aircraft without any fuel in it! This is the weight of the aircraft and everything in it, all the fittings, all the payload etc. For the standard passenger -400 it is 242,671kg (535,000lb).

So we can see that at MZFW (ie with maximum payload) the maximum fuel that the aircraft can get airborne with is (MTWA minus MZFW = 394,625 minus 242,671) 151,954kg, which is less than the capacity of the aircraft's fuel tanks (approximately 173,000kg). If the aircraft is filled with fuel, the Zero Fuel Weight cannot be above (394,625 minus 173,000) 221,625kg, thereby restricting the payload. There is therefore a choice between maximum payload or maximum range.

We can also see that the maximum payload that can be lifted will be somewhere in the region of (242,671kg minus 180,000kg) 62,671kg, depending on the actual empty weight of the particular aircraft.

There is also a certified Maximum Landing Weight (MLW); for the standard passenger -400 this is 285,763kg (630,000lb). The aircraft cannot be planned to land above this weight; however, in an emergency it can be landed at any weight up to MTWA (but this would require a heavyweight landing check by the engineers).

For each and every take-off there will also be a Performance Limited Take-Off Weight. This is calculated by the pilots as part of their pre-flight preparation and takes account of such things as the runway length and condition, airfield altitude, obstacles that the aircraft has to climb over, any deferred defects with the aircraft and the environmental conditions (air temperature, air pressure, wind strength and direction). For the purpose of this calculation it is also assumed that a critical engine fails (for the 747 that is an outboard engine) at a point during the take-off where it is possible to either stop the aircraft on the remaining runway, or continue and get airborne without hitting anything (missing it by a set margin). This

point is defined as a speed, known as V1. The Performance Limited Take-Off Weight can be greater than the structural MTWA (ie on a long runway at sea level in flat terrain on a cold, dry day) or significantly less (high altitude airport surrounded by mountains with a short runway on a hot day with heavy rain).

This calculation is done for *every* take-off and whichever is the more limiting, the Performance Limited Take-Off Weight or the MTWA, is the maximum weight at which you can get airborne for that particular flight. This is known as the Regulated Take-Off Weight (RTOW). The aircraft is not allowed to take off at a weight greater than the RTOW. So if the planned take-off weight exceeds the RTOW the choices are either to offload some payload so that enough fuel can be taken on board to get the flight to its planned destination, or take all of the payload and a reduced fuel load and stop en route to refuel. However, the RTOW is determined by a wide range of variables, so another option might be to wait for the environmental conditions to change (perhaps with the cooler air temperature at night) or to use a different runway.

Every operator has its own seating configuration; in fact operators may have more than one for their fleet if their route structure dictates that requirement. They may also change over time as market conditions alter, and the aircraft was designed with this in mind. A typical business route will see a lot of First- and Business-class seats and relatively few Economy ones, so the total seating capacity could be around 350, whereas a leisure route will have few if any First- and Business-class seats and may be entirely Economy with around 500 seats. However, the aircraft is actually certified for a total of 678 passengers (plus any number of infants under 2 years old).

Initial preparation

Calculating the RTOW is just a part of the pre-flight preparation performed by the pilots. The pilots will also be checking the weather forecasts and the NOTAMs (Notices to AirMen) looking for any information that may be relevant to the flight. They will have been provided with a flight plan, which will show the planned route, the expected fuel burn at the planned weight and the minimum fuel required.

RIGHT Keeping the hot sun out of the cockpit by using reflective sun shades. This is a Cathay Pacific 747-400 at Hong Kong in June 2012. *(Ian Black)*

They will also have a look at the Tech Log to see if there are any defects that they need to be aware of. Having done all that they will then decide how much fuel to load.

Having established the RTOW and decided on the fuel load, the planned take-off weight will be known. The same methods (AFM, tabulated data or TODC) can be used to determine the power setting and speeds for the planned take-off weight for the particular runway being used.

At the aircraft

The aircraft starts to come alive once either the ground power is plugged in and selected ON or the Auxiliary Power Unit (APU) is started and its generators are selected ON.

For a normal revenue flight, from entering the flight deck it takes about an hour to be ready for pushback and engine start. During this time there is a lot to do: the initial preparation of the flight deck, the loading of the Flight Management Computer (FMC), the calculation of the Take-Off Performance, the exterior inspection (as with any aircraft), the receipt and checking of the load sheet and the subsequent loading of that data into the FMC, the Departure Briefing, and finally a chat with the passengers to welcome them onboard.

The initial preparation requires a scan of all the panels to ensure that all the switches are in the required position for flight. Other than switching on the three IRS, there are actually very few selections to be made.

Loading the FMC

This is the meat of the pre-flight preparation and takes the most time. If you start at the beginning the FMC leads you through the set-up, by following the prompt in the bottom right-hand corner of the CDU screen.

RIGHT Ground servicing the aircraft on turnaround. *(Boeing)*

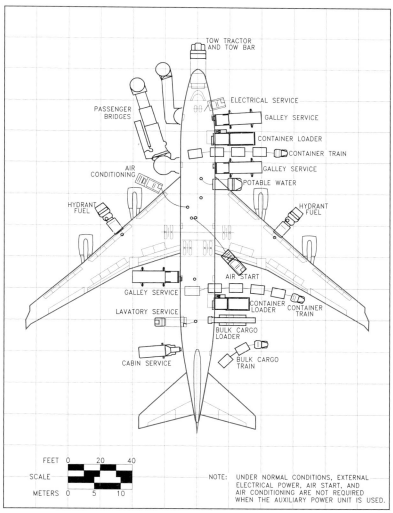

NOTE: UNDER NORMAL CONDITIONS, EXTERNAL ELECTRICAL POWER, AIR START, AND AIR CONDITIONING ARE NOT REQUIRED WHEN THE AUXILIARY POWER UNIT IS USED.

FROM BOEING 747 CLASSIC TO -400 AND BACK AGAIN

Captain Ian Frow

It was probably in 1987 that I was invited to become one of the 'founding five' training captains on the yet-to-enter-service Boeing 747-400 of British Airways. During the next two years we all became increasingly involved with reading the proposed manuals for this very different 747. The manuals had many new and curious abbreviations and, at times, it was really quite difficult to understand from them how this oh-so-clever machine worked.

On visits to Seattle we also became involved in some minor aspects of the flight-deck design. I distinctly remember helping to decide where the pilots' torches should be stowed on the flight deck and also the exact positioning of the

TOGA switches on the thrust levers. Perhaps the most bizarre exercise I experienced was a day spent in an engineering simulator in Seattle flying endless instrument circuits, whilst the vertical strip ASI was altered between 'high speed up' and 'high speed down'. The glass cockpit display philosophy had not quite settled down then!

In late 1988 we did the technical course and flew the simulator, which at that time was not especially reliable. It had an alarming tendency to dive uncontrollably towards the very realistic visual display of the ground, leading to stressed nerves and raised blood pressure. Our instructors were Boeing test pilots who flew the simulator like a dream, but not all were born instructors. One had a memorable way of describing how this very electronic aeroplane worked.

'You see it has all these electrons a-hanging on to each other'. Here he linked his two hands by just the tips of his fingers, 'And if those little critters let go of each other', his hands flew dramatically apart, 'man you got problems!'

The entry into service was delayed and for six months we had a strange half-life instructing on the Classic on the simulator/base flying/route training, whilst also keeping our hands in (and sometimes instructing) on the now improved -400 simulator.

Eventually in summer 1989 it was decided that we 'pioneers' would have to do large parts of the course all over again; but this time in the simulator we instructors were checking and instructing each other. It was an odd time but eventually we all went out to Seattle to meet the -400 in the flesh for the first time. One novelty for us Classic men was the need to do one's own outside check (a task carried out on the Classic by the trusty flight engineers, who really knew what they were looking for).

Base flying out of Seattle was carried out using the massive long runway just over the Rockies at Moses Lake. If two aircraft were in the circuit at the same time and the wind was negligible we flew 'banjo' circuits. Following the 'touch and go' and at about 300ft a 90° turn was commenced and the heading held for 30 seconds, followed by a turn on to the downwind and base legs for an approach, 'touch and go' on to the opposite direction runway. Thus once established on approach there was the interesting spectacle of the other aircraft flying towards you before entering his 'banjo' turn. From the air it looked extraordinary, from the ground it must have been awesome, but it certainly got the circuits done quickly.

The initial few months of passenger carrying route flying was trying to say the least. Again we training captains were initially route training and checking each other, an exercise complicated by the fact that one of the first three aircraft developed just about every curious electronic fault that the design could produce (there was a rumour, no doubt untrue, that the wiring looms had become loose in places). Consequently it was quite rare to arrive home on the scheduled day or even the same week. All these problems did mean that, perforce, we became remarkably knowledgeable about the aircraft's systems.

Compared to the Classic the -400 was a very much more automatic aeroplane and Boeing made much of this feature in their training. Some older heads soon realised that, basically, the -400 flew and could be operated very much like the Classic. They felt that perhaps the course should have emphasised this, whilst showing how the automatics were a very desirable add-on, which immensely enhanced the operation. We felt that in some ways they put the cart before the horse.

The other big difference was the absence of the 'third head' – the flight engineer. Many of us initially felt naked without them. Not only were they a source of deep technical and engineering knowledge; in the past the third brain and third pair of eyes had saved many an incautious pilot from error – especially at the end of a deeply fatiguing long-haul flight. (Once ultra-long-haul flights commenced on the -400 there were of course additional pilots – and pairs of eyes – but they never had the flight engineers' ability to find a cheap breakfast).

Then it was off to the 'Retirement Job' with Virgin Atlantic. They were not able to use any of my hard-won -400 knowledge as they didn't have any -400s at the time and put me back on the Classic. Initially I was in the right-hand seat, but within a year back in the left-hand seat as a training captain. Those Virgin Classics were an amazingly mixed bunch of aeroplanes, sometimes quite different to each other and with each aircraft having a real character, something the -400 aeroplanes never developed. The star was 'Sir Freddie Laker' a -100 series Classic with not too much power but the sweetest of handling characteristics. And of course sitting behind one again was a flight engineer, whose technical expertise really came into its own when dealing with problems on this mix of second-hand aeroplanes.

So how would I compare the Classic 747 and the 747-400? Well, especially when handling Virgin's Classics, the nearest comparison would be the difference between driving a steam railway engine like the *Flying Scotsman* or a fully automated Eurostar express. It was a different world.

ABOVE **Passenger air-bridge.** *(Kickers/iStock)*

Start with the INIT REF (Initial Reference) button and go to the IDENT (identification) page. Here you can check that the navigation database is in date (it is updated every four weeks), and also that the correct performance degradation is programmed. As aircraft age they acquire wrinkles and dents that affect their aerodynamic performance to a small degree. This is most noticeable in the amount of fuel the aircraft burns, so it is possible to compensate for this in the FMC by adding a percentage correction factor. Next go to POS (position). It

RIGHT **Air stairs.**

is very important to input the correct position to enable the IRUs to align correctly. The aircraft's exact position can be determined from the aerodrome chart.

Next go to the RTE (route) button and on RTE page 1 input the departure and arrival airports, using ICAO four-letter designators (such as EGLL for London–Heathrow, KLAX for Los Angeles), and the flight number. It is very important to enter the correct flight number, as the automatic reporting functions will use this. You can now go to the DEP/ARR (departure and arrival) button and input the planned runway and Standard Instrument Departure (SID). Go back to RTE page 2 and input the route using waypoint identifiers and airway identifiers. These may be on the flight plan or can be obtained from airways charts. Once the route is loaded go back to DEP/ARR and enter the planned Standard Terminal Approach Route (STAR) and the expected instrument approach, then go back to RTE. Select Activate and press the Execute button; the route is now loaded and active so will appear as a magenta line on the ND.

Next go to back to INIT REF and select the PERF (performance) page. Enter the

ABOVE Baggage loading into G-VROS in Orlando.

LEFT Baggage loading. *(Matthias Clausen/iStock)*

LEFT Turnaround at Tokyo International. *(iStock)*

planned Cruise Altitude, Cost Index (CI) and Reserve Fuel. The Cost Index is calculated by the operator and is the ratio of hourly cost, excluding fuel, to the cost of fuel. The FMC calculates its ECON (economy) speeds based on the CI. If the cost of fuel goes up the CI goes down, and the lower the CI the lower the speeds, so a CI of 0 will result in the lowest fuel burn but the longest flight time. Reserve fuel is the minimum amount of fuel needed to get to the planned alternate airfield if a landing at the destination is not possible.

Next go to THRUST LIM (thrust limit); there is nothing to enter here yet, but the indicated N1 (GE) or EPR (P&W and RR) can be checked, then go to TAKE OFF, enter the take-off flap setting (10 or 20), acceleration height, engine out acceleration height and thrust reduction (there are several options, it can be an altitude or a flap setting).

A quick look at PROG (progress) page 1 will show the total route distance. This can be used as a gross error check against the total distance on the flight plan.

The next action is to go to the LEGS page and check that all the tracks and distances between waypoints agree with the flight plan or chart. This is a double-check to make sure that no mistakes have been made with loading the route. It is also a check that information in the FMC's database is correct.

Now on the LEGS page go to RTE DATA (route data) and input wind and temperature information for some or all waypoints (this information can also be uploaded automatically via the ACARS).

Finally go to NAV RAD (navigation radios) and select any radio navigation beacons that may be required for the departure. The FMC will automatically tune VOR beacons, but they can also be selected manually and the ADFs have to be selected manually.

The initial preparation of the FMC is now complete.

Briefing

Prior to pushback a pre-departure briefing will normally be given by the pilot actually doing the flying after pushback. This serves several functions: it makes sure that the other pilot(s) know what the flying pilot is planning

to do; it is a chance to double-check that the FMC is correctly loaded for the expected departure; and it also refreshes their memories of the actions to be taken in the event of any emergencies.

Load sheet

A load sheet is provided for each flight. This details the various weights of the aircraft (actual Zero Fuel Weight, actual Take-Off Weight, expected Landing Weight), the loading of the aircraft and therefore its centre of gravity. From this the position to set the stabiliser is calculated so it is trimmed for take-off. The load sheet can be transmitted electronically via the ACARS, or done manually on a large sheet of paper.

Final preparation

Having received the load sheet, and once fuelling is complete, the aircraft's Zero Fuel Weight can be entered on the CDU PERF INIT (performance initialisation) page. As the FMC knows the fuel load it will calculate and display the aircraft's actual weight.

Next go to THRUST LIM (thrust limit) and select the take-off thrust that was worked out earlier as part of the take-off performance calculation. Then go to TAKE OFF REF (take-off reference) and enter the calculated speeds; V1 (decision speed), VR (rotation speed) and V2 (take-off initial climb speed), and the calculated stabiliser position.

On the MCP arm the LNAV and VNAV modes; the annunciations will be TOGA and TOGA for pitch and roll, with LNAV and VNAV showing armed in white. The autothrottle annunciation will be blank. Complete the initial part of the Before Start checklist and the aircraft is now ready for departure.

Engine start and pushback

Once clearance to start engines has been received, select the fuel pumps ON, select the No 4 hydraulic system Demand Pump to AUX. This pressurises the hydraulic brake accumulator, ensuring there is brake pressure during pushback prior to starting engines. Switch the other three hydraulic systems Demand Pumps to AUTO. Turn off two air conditioning packs and complete the Before Start checklist.

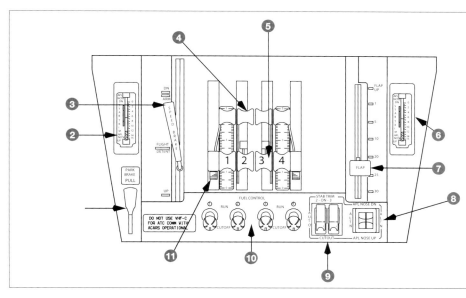

Caption and key to come

1 Parking brake lever
2 Captain's stabiliser trim indicator
3 Speed-brake lever
4 Reverse thrust levers
5 Thrust levers
6 First Officer's stabiliser trim indicator
7 Flap lever
8 Alternate stabiliser trim switches
9 Stabiliser trim cut-out switches
10 Fuel control switches
11 Autothrottle disconnect switches

The engines can be started in any order but No 4 is normally started first as the No 4 hydraulic system is the primary power source for the main wheel brakes.

Starting is achieved by pulling the respective engine's start switch in the overhead panel and then selecting its fuel control switch to RUN. Each start is controlled by the engine's EEC, normally using the Autostart system. Using the Autostart system it is possible to start two engines simultaneously.

If the aircraft requires a pushback from its

LEFT Engine start. *(Ian Black)*

BELOW De-icing a 747. *(Ian Black)*

parking stand, the parking brake is released before the pushback commences. The engines can be started during the push. Once the pushback is complete the parking brake is reset before the tug and tow bar are disconnected. There are a few actions to be performed after starting engines, such as switching off the APU, selecting the No 4 hydraulic Demand Pump to AUTO, switching the two air conditioning packs back on, turning on the aft cargo heating and most importantly selecting the take-off flap setting.

Complete the After Start checklist.

The following is a brief résumé on flying the aircraft. Unfortunately, space does not permit a more detailed discussion, or a discussion of the capabilities of the FMC (which could easily fill another book).

Taxying

Release the parking brake and the aircraft will start moving at idle power. At heavy weights it will require an increase in power to get it moving but power can be reduced back to idle once under way. The aircraft is steered on the ground using the nose wheel steering, controlled by a pair of tillers, one on each sidewall for each pilot. The combination of nose wheel and body gear steering allows a 180° turn on a 153ft-wide pavement. When you consider that the aircraft's wingspan is 213ft and the fuselage is 232ft long, that is impressive!

LEFT Line up –
entering the runway.
(Karl Drage)

BELOW View from
the top. The height of
the 747's flight deck
above the ground
gives the pilots an
almost panoramic
field of vision.
(Jorg Hackemann/
Shutterstock.com)

The view from the flight deck is excellent, except for close-in obstacles, which cannot be seen owing to the height from the ground – during the pushback you cannot see the tug! It is possible to see 90° either side of the nose from either seat and to get a view back past the wingtip on the side you are sitting. During taxying, check the flying controls for full and free movement, monitoring the movement of the surfaces on the EICAS status page. All the aircraft systems are being continuously monitored and will signal any problems by displaying an EICAS message. This means that the pilots can concentrate on taxying the aircraft and keeping a good lookout. It is a big aircraft!

Pre-take-off
Complete the Before Take-Off checklist.

Take-off
When cleared to enter the runway, taxi into position and line up on the runway centreline. When cleared for take-off slowly advance the power to 70% N1 (GE), 1.1 EPR (P&W), or 1.2 EPR (RR) to check for symmetrical engine spool-up. Once that is confirmed press the TO/GA switches. This engages the autothrottle in THR REF mode, advancing the thrust levers to the pre-computed take-off power. The PFD annunciations will be THR REF (for the autothrottle), TOGA (for roll) and TOGA (for pitch). At 65kt Indicated Air Speed (IAS)

ABOVE In the queue awaiting clearance from the tower. *(Ian Black)*

RIGHT Take-off San Francisco style. B-HKU, 6 September 2010. *(Chris Wood)*

BELOW The view from the cockpit on rotation, looking down on to Heathrow's Terminal 4. *(Ian Black)*

the autothrottle disengages (the annunciation changes to HOLD), at 80kt EICAS warnings are inhibited and at 85kt the RTO function of the autobrakes is armed.

Keep the aircraft straight using the rudder pedals. Initially you will be using the nose wheel steering, and then the rudders as airspeed builds and they become effective.

On reaching the rotate speed, VR, smoothly and continuously pull back the control column at a rate of 2.5° degrees of pitch per second to achieve a pitch attitude of approximately 15° nose-up. Keep the wings level. The aircraft starts to fly at around 10° and should stabilise at the initial climb speed of V2 plus 10kts. V2 varies from around 127kts for an empty aeroplane with minimum fuel (around 200 tons) to 182kt at Maximum Take-Off Weight (normally just under 400 tons – although the Extended Range version has a Maximum Take-Off Weight of well over 400 tons!). Once the aircraft is airborne the landing gear is retracted by moving the landing gear lever to the UP position.

At 50ft the LNAV mode activates; passing 400ft the VNAV mode activates, the autothrottle re-engages and the EICAS messages are uninhibited. The annunciations will now be THR REF, LNAV and VNAV. The autopilot can be engaged from 250ft above the ground.

Take-off

1 Rotate.

2 Gear up. The gear doors open first.

3 and 4 Gear retracts under hydraulic pressure.

5 Gear retracted, doors closed.

being retracted and it accelerates, but this can easily be countered with stabiliser trim. When the flaps are retracted the outboard ailerons are locked, reducing the roll rate.

Complete the After Take-Off checklist.

Accelerate to V2 plus 100kt for the initial climb. At some point the autothrottle automatically selects Climb Power. This can be at a set altitude or at a particular flap setting as programmed in the FMC. Passing 10,000ft (most countries have a speed limit below this height) the aircraft will accelerate to its FMC-computed optimum climb speed and will maintain this speed up to the Top Of Climb (TOC).

Climb

Maintain a speed of V2 plus 10kt up to the flap retraction altitude (as set in the FMC), then lower the nose to reduce the pitch to around 7.5°. The aircraft will continue to climb and as the speed increases retract the flaps in stages, at the appropriate minimum manoeuvring speeds, which are displayed on the PFD speed tape. At V2 plus 20kt Flap 10 is selected; at V2 plus 40kt Flap 5; at V2 plus 60kt Flap 1; and at V2 plus 80kt Flap Up. The aircraft does have a tendency to pitch nose-up as the flaps are

Cruise

The autopilot will capture the selected cruise altitude and the autothrottle will bring the thrust back to that necessary to maintain the cruise speed, typically around Mach.85 (which is around 490kt TAS). Normal cruising altitude varies with weight – as fuel is burnt and the aircraft weight reduces, the optimum altitude increases. On a long flight, having taken off near maximum take-off weight, you would expect to be at around 32,000ft initially and end up at around 38,000ft towards the end of the flight. Maximum speed (MMO) is M.90 and the fuel consumption is around 10 tons per hour.

Handling

The aircraft is very stable at all altitudes and speeds and can easily be hand-flown from take-off up to cruise altitude and through descent and landing. With four powerful under-slung engines, any power increase results in a pitch-up; power reduction results in a pitch-down.

Stalling

This is something not normally seen, except in a simulator. The 747 is remarkably benign in the stall and can easily be recovered. As the aircraft approaches the stall, buffet is felt giving a natural warning of the approaching stall (as well as stick shaker activation). At the stall the nose pitches down and any wing drop can be corrected with aileron. The recommended recovery technique is to unstall the aircraft first by lowering the nose and letting the airspeed increase, before applying power. Be cautious

RIGHT Climb power. (Ian Black)

ABOVE En route,
North Atlantic.

when applying power – because of the aircraft's tendency to pitch nose-up it can easily re-enter the stall.

Descent

Prior to descent complete the landing preparations. These include entering the flap setting (normally Flap 30) and the approach speed, referred to as VREF, in the FMC.

This will generate minimum flap manoeuvring speeds on the speed tape of the PFDs. The minimum altitude for the expected approach can be set on the PFDs as either a barometric altitude or a radio altitude.

Complete the Descent checklist.

Ideally descent is flown with idle power from cruise altitude. This gives a rate of descent of around 3,000ft per minute at 300kt IAS. For descent planning, allow 3 miles per 1,000ft plus 1 mile per 10kt speed reduction, and compensate for wind. Typically, therefore, descent from normal cruise, in the high 30,000s of feet, will require in excess of 100 miles. The FMC will calculate the Top Of Descent (TOD) point, and if a lower altitude is selected in the MCP altitude window prior to reaching it, the autopilot will initiate a descent at that point. The pitch annunciation will stay as VNAV PATH. If a lower altitude is selected after the TOD point, descent can be initiated by setting a lower altitude and pressing the altitude selector knob. The pitch annunciation will be VNAV SPD.

Prior to the approach complete the Approach checklist.

Approach

The aircraft is aerodynamically very clean and has a lot of residual thrust at idle power, so it takes a long time to slow down and uses up lots of airspace. As the speed reduces the flap is selected in stages in the same, but reverse, way to the initial climb.

The aim is to fly at or above the minimum manoeuvre speed for the flap setting, so with the flaps UP the minimum speed is VREF plus 80kt; with Flap 1 it is VREF plus 60; with Flap 5 VREF plus 40. The intermediate

LEFT Descent.
(TopFoto)

approach is flown with Flap 10 at a minimum speed of VREF plus 20kt. As you approach the Glideslope select Flap 20 and the landing gear DOWN. Arm the speed-brake once the landing gear is down. As the Glideslope is intercepted landing flap is selected, normally Flap 30, and speed reduced to a landing speed of VREF plus the wind additive. There is a noticeable nose-down pitch when this stage of flap is selected.

Complete the Landing checklist.

The aircraft is capable of several different

types of instrument approach, including an Instrument Landing System (ILS) approach and a GPS-based RNAV (aRea Navigation) approach. They can be flown automatically using the AFDS or be hand-flown. With all of them a 3° Glideslope is normally flown, using whatever aids are available. The aim is to be fully stabilised in the landing configuration by 1,000ft above the ground for a straight-in approach. At normal landing weights a rate of descent of around 700ft per minute equates to a 3° slope, and a power setting of around 65% N1 is required.

A visual approach can also be flown. If joining the visual circuit, fly downwind and pass abeam the landing threshold with Flap 10 selected at an appropriate speed. Continue flying downwind for 15secs per 500ft of altitude, as you approach the end of the downwind leg select Flap 20, the landing gear DOWN and start a descent and turn. Select FLAP 30 and aim to be rolled out on finals by an absolute minimum of 500ft agl.

The approach speed (VREF) is a function of the aircraft's weight and flap setting. The easy

one to remember is that at a landing weight of 240 tons with Flap 30 selected the VREF is 140kt. This alters by roughly 1kt per 3 tons of weight and is 153kt at the normal maximum landing weight of 285.7 tons. If using Flap 25 the VREF is around 5kt higher for a given weight. The aim is to fly the final approach at VREF plus an addition for any wind. The wind addition is a minimum of 5kts, even with a tailwind, and up to half of the steady headwind component, plus all of the gusts, to a maximum of 20kt.

So, for example, if you are landing on Runway 36 and the wind is 360° at 15kt gusting 25, you must add 8kt for the headwind (half of the headwind component) plus 10kt for the gust, and you will fly your approach at VREF plus 18kt. If the wind was 090° at 15kt gusting 25, you would add 0 (headwind *component* is 0, it is all a crosswind) plus 10kt for the gust, so you would fly at VREF plus 10kt.

Landing

At 30ft radio altitude the flare is initiated and power slowly reduced to idle. If there is

a crosswind push the aircraft straight with the rudders and keep the wings level with the ailerons. The main wheels touch first, automatically initiating main wheel braking and deploying the speed-brakes. Reverse thrust is selected manually once the thrust levers have been moved to the closed position. Spoiler deployment helps to dump lift from the wings, putting weight on to the wheels to enhance braking, and also results in a nose-up pitching moment. Gently lower the nose wheel and keep the aircraft straight using the rudder

ABOVE Over the threshold. *(Ian Black)*

BELOW Landing, Schiphol. *(Michael Winston Rosa/ Shutterstock.com)*

pedals. Gentle forward pressure on the control column once the nose wheel is lowered also aids this. The autobrakes will bring the aircraft to a stop.

Automatic landing

The aircraft is certified for automatic landings, which are primarily used in conditions of low visibility. To do this it requires both the aircraft systems and the specified ground equipment to be working correctly. It can only be done with an Instrument Landing System (ILS) that is certified for the category of landing being attempted.

When APP (Approach) mode is selected with an autopilot engaged, the other two autopilots are armed. Passing 1,500ft radio

altitude these two autopilots engage and the PFD annunciation LAND3 appears, and the ROLLOUT and FLARE modes are armed (with annunciations in white). At the same time the AC electrical busses are split so that each autopilot channel has a separate electrical supply, and the EICAS electrical synoptic display is inhibited.

Passing 500ft radio altitude the Runway Align sub-mode is engaged. This starts to line the aircraft up with the runway, using the rudder channel, if there is any crosswind, so that it is effectively flying cross-controlled to keep it straight.

At approximately 50ft radio altitude the aircraft commences a flare and at 25ft radio altitude the autothrottle retards the thrust levers to idle.

At 5ft the wings are levelled, on touchdown the speed-brakes deploy, autobraking commences (assuming both have been armed) and the nose is lowered. The only action required of the pilots is to select reverse thrust. The rollout mode will track the runway centreline and the autobrakes will bring the aircraft to a halt.

Shutdown

Having cleared the runway (remember there is a *lot* of aeroplane behind you) and established the taxi route to the parking gate, the after landing actions such as retracting the flaps and retracting the speedbrakes can be completed. Having taxied the aircraft to its parking spot, starting the APU on the way there, all that is required before shutting the engines down is to set the parking brake, select the No 4 Demand Pump to AUX (to ensure hydraulic pressure for the parking brake through the accumulator) and select the APU generators ON. Moving the fuel control switches to CUT OFF shuts off the fuel supply and the engines wind down. The only engine cooling requirement prior to shutting them down is to run the engines for three minutes after cancelling reverse thrust.

Complete the Shutdown checklist.

The 747 is very much a pilot's aeroplane, despite the automation. By disengaging the autopilot and the autothrottle (just two clicks) it becomes like any other aeroplane and is easy to fly – a true testament to Joe Sutter and his team. They got it right!

LEFT Landing at Las Vegas, Virgin's G-VXLG.

LEFT Flare.

LEFT Reverse thrust selected, speed-brakes deployed.

FLIGHT ENGINEER ON THE 747-100

Senior Engineer Officer Ralph Chadwick

Until the arrival of the -400 in the late 1980s with its high level of automation, all 747s had a three-man flight crew – two pilots and a flight engineer. Ralph Chadwick was an engineer on the Boeing 747-100 with British Airways.

Ralph joined the Royal Air Force in 1951 and was trained as an Engine Fitter. He had postings to various RAF stations in the UK before finally deciding on an overseas posting to Malaya, serving at RAF Butterworth and finally at RAF Changi, where he was allocated to the Staging Aircraft Servicing Section. Ralph serviced all kinds of aircraft that passed through his hands, including the Avro York and Handley Page Hermes, which at that time were engaged in trooping contracts. On leaving the RAF he went to work for Hunting-Clan at Heathrow before gravitating to BOAC, where he worked on the DC-7C and Argonaut and thence to Flight Engineering, the Britannia 102, VC10 – and ultimately the Boeing 747.

Aircraft development depends on the engine power available at the time, which is the limiting factor in producing large airliners. This has varied with time, from the large piston engines that were fitted to aircraft like the Boeing Stratocruiser in the late 1940s up to the fan engines used to power the large jets of today.

For instance, the Convair company in the USA produced the XC-99 in 1949. Not only was it a 'Jumbo Transport', but it also had a full-length double-deck (shades of the Airbus A380); however, it needed six of Pratt & Whitney's 28-cylinder 4360 (3,400hp each) engines to power it. In the UK the Bristol Aeroplane Company produced the Brabazon as a potential transatlantic aircraft. Yet, this needed no less than eight Bristol engines, coupled in pairs, driving four sets of contra-rotating propellers (a flight engineer's paradise, but a ground engineer's nightmare).

To put these aircraft in context the wing-span of a 747 was 195ft, the Brabazon 225ft and the XC-99 was 230ft. It took until the late 1960s and the development of large fan

ABOVE Senior Engineer Officer Ralph Chadwick at his engineer's panel on a British Airways 747-100. *(Ralph Chadwick)*

BELOW British Airways 747-136, G-AWNB, *Llangorse Lake*, holding for take-off clearance at London–Heathrow in 1989. *(Jonathan Falconer)*

engines before the very large aircraft we know today became feasible. As an example the take-off thrust of each of the VC10's four Rolls-Royce Conway engines was 22,500lb, compared to the take-off thrust of the early 747 engine (JT9D) of 45,800lb and the Rolls-Royce RB211 engine, which delivered 51,500lb of thrust, although not always – but that's another story!

The first flight engineering position I held was on the Bristol Britannia 312, which was a challenge simply because of its miles of electrical systems – from electric throttles controlled by amplifiers, which frequently produced a failure condition, to the large selection of relays, inverters and so on all located in the electronics bays on either side of the freight holds. So the first thing you had to do before troubleshooting a system was to move all the baggage to one side, a tedious task to say the least.

From there I moved to the Vickers VC10 fleet, an aircraft that was a delight to fly, plenty of power available for take-off – and to my absolute delight a lovely flight engineer's position with your own set of throttles (sheer joy). This was not an afterthought as in the Britannia, but a beautifully designed well thought-out panel. Little did I think that the VC10 would see me out. It's still flying today and I'm not.

As the '60s came to a close I was offered a position as a flight engineer on the Concorde fleet by that wonderful engineers' engineer, Lou Bolton. It took me a long time to consider this offer – and who wouldn't want to fly the most glamorous aircraft in existence – but I declined since I was already licensed on both the Britannia and VC10. I was conscious of what the future held for me so I volunteered to fly the 747, which would take me into retirement and further if I wished to continue flying with another company.

The introduction of the 747 changed much in civil aviation, inasmuch as we flight engineers sat an Operations Course, which brought us up to the standard of a Commercial Pilot's Licence. We studied subjects such as meteorology, navigation and flight planning. Along the way we gained an R/T licence so we could handle radio calls to the company and others when required. We also held a limited engineering approval that made the aircraft completely independent if it was needed to fly into somewhere off the beaten track.

What training was necessary to fly the 747? The system was remarkably different to that for the VC10, where we did our training as engineers in a separate block to the pilots. It took approximately ten weeks then we came together with the pilots for our simulator periods and live flying training at Shannon. I have some wonderful memories from that period, for example flying parallel to the cliffs at Moah at low level. An experience never to be forgotten.

Our training on the 747 took a different form as we were with the pilots from day one, with a month in the ground school (to which all the flight engineers turned up their noses, since it was considered inadequate for such a large aircraft). But four weeks it was, then into the simulator to undertake a series of exercises to explore first the normal operation and then on to the normal failures that occur on all aircraft and the drills to overcome them, plus the emergencies. All eventualities were covered, from a stuck stabiliser to a double-engine failure on one side. This was the stage where the flight engineer became 'the commander of the checklist', from the end of the after-start checklist to the culmination

BELOW The flight deck of a Pan Am 747-100 on the airline's first 747 service from Paris to New York on 5 February 1970. (TopFoto)

of the after-landing and shut-down checklists
and all other times in between.

From a ground handling point of view,
during hands-on operation of the aeroplane
it was stressed that we had a vast amount of
thrust available to us. We had to be aware of
the dangers of blowing light aircraft away and
damaging ground installations during taxying.
All in all, one had to keep alert unless you
wanted an uncomfortable interview with the
Flight Manager on return to base.

New innovations for me and used on the
747 were the Auxiliary Power Unit (APU),
to ensure that the aircraft was independent
of external power and air sources, and of
course the Inertial Navigation System (INS).
This computer was an absolute joy since it
provided the answer to every navigational
problem ever thought of and provided
stabilisation for the flight instruments.

The first impression I had on walking out
to the aeroplane for the very first time was
how large and imposing a beast this machine
was. I felt a moment of pride as I thought of
tiny diminutive me being in charge of this very
large aircraft (from an engineering point of
view, of course). I well remember one of my
colleagues saying, 'Is this the forward freight
hold?' as we looked around the aircraft's wing
gear (the undercarriage unit itself was 11ft
high) and the size of the aperture meant to
accommodate it in flight.

From there the crew split and the pilots
went to the flight deck whilst the flight
engineer did an exhaustive outside check
of the aircraft, which largely consisted of a
condition check of the external structure and
a check on the tyre state. Was the fuel being
loaded? It took approximately 30 minutes to
pump in excess of 30,000gal of best JP1 fuel
into the tanks.

On entry to the flight deck the first
impression was how the flight engineer's station
was nearly a third longer than that of the VC10.
The panel was organised in such a way that
took one's eye seamlessly to the next item and
followed a natural 'flow'. This pre-flight scan
was accomplished without a checklist, which
was something I wasn't used to, but I quickly
became accustomed to the Boeing way; a little
mind-boggling nevertheless. The panel was very
easy to operate with each item falling to hand
as required.

The ergonomics of the 747 did not match
up when compared to those of the VC10,
where I had all of the engine controls and
instruments in front of me except the HP
cocks. All in all the early models of the 747
were a significant challenge to pilots and
engineers alike – the sheer scale of the
aircraft and the fact that on the ground 'the
office' was 30ft above the ground, 70ft when
overhead the runway threshold when landing
– it was a significant factor, which I was
quickly able to master. Crews used to say that
it was like trying to land a block of flats from
the upstairs toilet window!

Coping with the new technology engines
added to the difficulty of a successful flight.
The large fan at the front was effectively a
large ducted propeller. Early models of the
747 were fitted with the Pratt & Whitney
JT9D-3A, a twin-spool engine where the LP
spool drove the fan and the HP spool drove
the accessory gearboxes. To ensure engine
stability from idle power to take-off power
required a complicated set of air valves and
control systems, which bled surplus air to
atmosphere. Unhappily they didn't always
work, so engine starting became an art rather
than a science.

You needed to take great care when
starting since the engine would very easily

go into surge mode, which consisted of a large and rapid increase in Exhaust Gas Temperature (EGT) and very little rotation of the engine. This resulted in a burnt-out first-stage turbine (which for some reason was frowned upon by the management). The silent surge could occur during the transition from reverse thrust to forward thrust during landing.

The anecdote that follows illustrates the difficulties of handling the aircraft when fitted with the early JT9D engines. If there was a failure of an engine during take-off the aircraft would only climb once you had achieved the correct speed, so it would either go along or up – but not at the same time. It is said that an American aircraft taking off from Heathrow went beneath a glider and tug at 2,000ft over Booker airfield after suffering an engine failure.

The Rolls-Royce RB211 engine was fitted to later models of the 747 and was a different animal altogether, having three spools, although it still had to be watched. But having an engine with three spools, each running at their optimum speed, made the engine much more flexible and stable in operation and not subject to the problems of previous engines. When the RB211 was under development the company had hoped to use carbon-fibre blades as the first stage of the fan, but they proved inadequate to the task and were replaced by hollow titanium blades on entry into service. The carbon-fibre blades were fitted to the VC10 and subsequently developed a somewhat ragged appearance due to rain erosion, so that was the end of carbon fibre. On entry into service the RB211 developed 51,500lb of thrust compared to 45,800lb of the JT9D. The difference in thrust was significant, which greatly improved the aircraft's performance.

The 747 became the workhorse and major profit maker within BOAC flying an average of 14 hours a day; any airline that didn't get that kind of utilisation wasn't doing it properly. With thorough maintenance the 747 went through at least two life extension programmes.

The most abiding memory I have of my time on the 747 was the exposure to all kinds of weather conditions, whether it was trying to land at Chicago in a snowstorm, with different runways in use and many other

aircraft using the tower frequency, to landing in Fairbanks, Alaska, at a temperature of minus 55 ambient. The outside check had to be very brief! Conversely, landing in Jeddah with the temperature at plus 50° C you ended up with very thin blood and of course there was no beer at the end as a relaxation. My favourite trips were down to the Caribbean where the attitude was so laid back, the sun never ending and the girls in the main were quite outstandingly beautiful – from an observer's point of view, naturally.

In my 32 years at BA and 5 years with Virgin Atlantic you could say that I led a charmed life, having shut down 2 engines on the Britannia, 3 on VC10s and in excess of 20 on the 747, all for various problems.

In those 37 years I spent 14 as a CAA Authorised Examiner in both airlines, which included scheduled CAA test flights after a major overhaul where the test schedule involved exploring the lower end of the speed range such as stalls (highly exciting) to the upper end. These flights lasted 3½ hours. All in a day's work! What a pleasure it was to meet such accomplished pilots as the CAA's Chief Certification Pilot, D.P. Davis, who didn't say much, but if the test schedule called for 310kt and a 50ft/min rate of climb that`s what you got; a remarkable man.

In my flying career of 37 years I logged a total of 27,000 flying hours. Would I do it all again? In a heartbeat.

ABOVE 747 flight engineer's panel on a 200 freighter. The panel layout varied between different airlines and models of the 747. *(Bas Tolsma/ TheDutchAviator.com)*

Chapter Six

Maintaining the 747

Large commercial aircraft like the 747 have a structured maintenance schedule, laid down by the manufacturer and approved by the operator's regulatory authority (such as EASA in Europe, or the FAA in the USA). Maintenance actions or inspections are generally done on either a calendar interval or a flying hour interval.

OPPOSITE A Lufthansa -400 undergoes maintenance at Lufthansa Technik at Rhein Main International Airport at Frankfurt in Germany. *(Frank Rumpenhorst/PA Images)*

Turnaround

The most basic check is a Turnaround Service which, as its name suggests, is done on every flight turnaround. This includes such items as checking the engine oil and hydraulic fluid levels, tyre pressures, refuelling the aircraft and a visual inspection to check for any signs of damage or wear.

Refuelling

The aircraft has a pressure refuelling point in each wing. These are behind closed

panels on the underside of the wing leading edge, between the engines. Each refuelling point has two receptacles, so two hoses can be used simultaneously on each side, and the panel in the left wing has a placard with detailed refuelling instructions. The system can refuel at a rate of 7,570 litres a minute, so with a total capacity of 216,825 litres it can take 30 minutes to fill an empty aircraft. (In normal operations the aircraft is normally fuelled from the left side only, so it takes considerably longer.)

Before refuelling is commenced, the fuel bowser must be bonded to the aircraft to equalise any static electricity, and minimise the risk of it igniting fuel vapour.

The refuelling system has an automatic function where the required fuel load is preselected. The system distributes the fuel correctly and shuts the aircraft's refuelling valves to stop the process once the required fuel is loaded. Alternatively the aircraft can be refuelled manually. With this method the amount of fuel going into each individual tank can be controlled.

It is also possible to gravity refuel the aircraft using overwing fuelling ports. Main tanks 1 to 4 can be filled this way, and the fuel can then be transferred to the reserve and centre tanks.

If any tank quantity is not showing, the quantity can be checked using the fuel measuring stick – release the stick, take a reading, note the aircraft attitude using the inclinometers in the left-hand body gear wheel

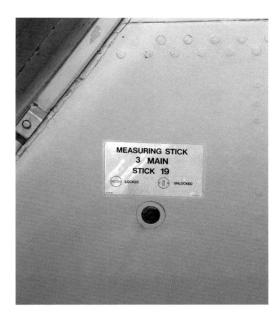

well and refer to the fuelling manual to calculate the quantity.

Water servicing

To fill the water tanks, open the water service panel and connect a water supply hose to the fill fitting. Pull the fill and overflow valve handle to the fill position. When sufficient water is loaded close the fill and overflow valve handle. When the tanks are full water will flow from the drain port. Close the water service panel (the panel will not close if the handles are not in the closed position).

To drain the water system, open the water service panel and the drain service panel. Connect drain hoses to the drain ports and pull both drain valve handles.

Waste servicing

The waste system is serviced at the waste service panel which is on the underside of the fuselage towards to rear. Specialist vehicles are used with pipes that have fittings to attach to the aircraft, and tanks to store the removed effluent.

Opening the engine cowls (GE engine)

To open the cowls start at the front, with the fan cowls, and work backwards. To open the fan cowls, first release the latches at the bottom, then engage the forward hold-open rod followed by the aft hold-open rod. Ensure that the rods are fully extended and locked (the red unlocked band is not visible), as the cowls are heavy and can cause damage and injury if not properly locked.

To open the thrust reverser cowls rotate the upper and lower latch handles to release the thrust reverser latch ring assemblies. It is now possible to open the thrust reverser. This is done using thrust reverser opening actuators, which are mounted on either side of the strut,

FAR LEFT Access for the fuel measuring stick to check tank contents. *(Chris Wood)*

BOTTOM The engines fitted to the 747-400 are incredibly reliable and often remain on the wing for years without needing to be removed for overhaul. Very occasionally, however, they do need changing and one way to deliver a spare engine for a 747 is to strap it to another aircraft and fly it there. This was common practice with the Classic 747s but is rarely used with the -400. VH-OJQ of QANTAS suffered a problem with one of its Rolls-Royce engines while climbing out of Singapore on 17 December 2009, which resulted in a return to Singapore. It is seen here returning to Sydney a few days later with the inoperative engine attached to the wing, inboard of the No 2 engine. *(Richard Goodman)*

and are hydraulically operated, either by hand or an electric motor.

To open the core cowls, first open the fan cowls and the thrust reversers. The core cowls have three latches along the bottom, open these and use the hold-open rods to hold the cowls open.

Closing the cowls should be done in the reverse, core cowls first, then the thrust reversers and cowls and finally the fan cowls.

Oil change (GE engine)

An oil change is rarely, if ever, needed, although the oil levels are checked after each flight and topped up as required. The oil tank has a filler cap, a sight gauge, a pressure fill port and an overfill port, and there is an access door in the right-hand side of the fan cowl for servicing the engine oil system. Wait at least 5mins after the engine has been shut down to allow any residual pressure in the system to dissipate, but no more than 30mins, to check the level. This can be done either on the secondary EICAS ENG page or the sight gauge on the oil tank. Oil can be added manually or under pressure. If adding manually the tank is full when oil spills over into the scupper; if under pressure it is full when it flows out of the overfill line.

When servicing the oil check for a smell of fuel. If it is present it would suggest a leak in either the fuel/oil heat exchanger or the servo fuel heater, so these should be changed and the oil system drained and flushed.

Opening the landing gear doors

The hydraulically operated landing gear doors can be opened on the ground for access to the various components located in the wheel wells. The main gear doors are opened by using two ground door release handles, one either side, which are near the aft inboard corner of the body gear wheel wells. To open the doors ensure that the area under them is clear, then press the release lever and move the handle to the down position.

To close the doors ensure that the number 1 and 4 hydraulic systems are pressurised and the area around the doors is clear. Operate the unlock lever and move the handle to the closed position. If the doors are not closed before flight, the unlock lever is engaged and the handle moved to the closed position as the gear retracts.

The hydraulically operated nose gear doors are opened by a separate ground door release handle. When operated this activates a nose gear door safety valve which unlocks the door actuator locks and allows the doors to fall open. To close the doors ensure the area is clear, that the No 1 hydraulic system is pressurised and move the handle to the close position. If the doors are not closed before flight the handle is moved to the closed position as the gear retracts.

Aircraft wash

Washing aircraft is an important maintenance function, as a build-up of dirt will impair the aircraft's performance, resulting in increased fuel burn and, therefore, cost. Dirty aircraft do not look good to the customers either. Lufthansa, for example, washes its 747-400s every 90 days.

The first phase is to apply a cleaning agent using spray lances. This is left on the aircraft for a short while before swabs are used to remove the worst of the dirt. Next, the aircraft is hosed down to wash all the dirt and cleaning agent away, and then the final phase is a clear water shower. A complete cleaning cycle takes 8hrs and uses around 10,000 litres of water.

BELOW Body gear hydraulically operated doors in the open position, which allows access to the wheel wells for maintenance. *(Chris Wood)*

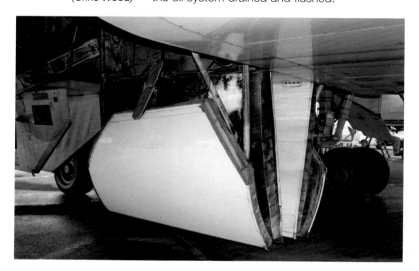

RIGHT Ready for maintenance, a -400F inside the giant hangar at Lufthansa Technik. *(Kickers/iStock)*

Routine maintenance

A maintenance schedule will be drawn up by the aircraft operator, based on guidelines from Boeing and approved by the operator's regulatory authority. This means that individual operators may work to different maintenance schedules. These are known as Continuous Airworthiness Maintenance Programmes (CAMP) and include routine and detailed inspections.

The detailed inspections are usually referred to as specific checks, starting with an A check. Boeing's baseline figure for an A check for the 747-400 is every 600 flying hours, although an operator's actual periodicity may be different. Assuming an average utilisation of 15 flying hours a day, an A check would be performed roughly every 40 days. It can usually be accomplished within 48hrs.

The next level of inspection is the B check. Although typically done roughly every six months, Boeing does not quote a periodicity for a B check for the 747-400. Normal practice is for the B check items to be completed during the A checks.

The next level is the C check. This is a major inspection and Boeing's baseline figure is every 7,500 flying hours or every 18 months. This is normally carried out in a hangar and can take up to two weeks to complete.

The most comprehensive check is a D check. Boeing's baseline figure for this is 6 years for the first one, then 8 years, then another 8 years, and then 6 six years. The D check involves dismantling much of the aircraft as well as a complete paint strip to aid the inspection process. A D check can take up to two months so is a very expensive operation, with costs potentially running into millions of dollars. Consequently it requires careful planning, both in terms of arranging for the aircraft to be out of service and arranging an overhaul location. As aircraft age and their values diminish, an impending D check can see an aircraft retired or put up for sale.

ABOVE Lufthansa Technik engineers conduct a visual inspection of the fuselage exterior of one of their -400s, D-ABTC, at Frankfurt. *(Boris Roessler/PA Images)*

BELOW The underside of the wing is inspected. *(Andrea Krause/iStock)*

What a D Check involves

A D check takes between 15,000 and 35,000hrs of labour, and can put an aircraft out of service for 30 days, or more. The total cost averages between US$1 million and US$2 million (at 1999 prices). 'A typical D check is 70% labour and 30% material,' said Hal Chrisman of The Canaan Group, an aerospace management consulting firm. Of course, some of that cost is included in your airline ticket.

Once the aircraft is parked inside the hangar – a huge complex of aircraft service areas, support shops, and warehouses – the maintenance team goes to work. Worktables, platforms and scaffolds are rolled into position for access to otherwise unreachable areas of the aircraft. Seats, floors, walls, ceiling panels, galleys, lavatories, and other equipment are opened or removed from the aircraft to permit close inspection. The aircraft is essentially gutted.

Following step-by-step instructions, workers examine the aircraft for signs of metal cracks and corrosion. Whole sections of the aircraft's

BELOW This aircraft has been stripped of its paint to aid the inspection process. The nose gear doors are open and a technician can be seen working inside the nose gear bay. *(Xin Zhu/iStock)*

Wheel change

With sixteen main wheels and two nose wheels, tyre changes on a 747 are a regular occurrence. Although it is usually the tyre that needs changing, the complete wheel still has to be replaced and sent to the workshop for the tyre to be removed. There are several reasons for changing the wheels, the most common being that the tyre has worn to its limit. Like all maintenance activities there is a clearly laid down procedure in the Aircraft Maintenance Manual to be followed.

A main wheel for a 747 with its tyre weighs around 130kg and can be changed by two mechanics in around 45mins. Cathay Pacific estimates that its 747 tyres last for 1,500 landings and in that time they will have been retreaded up to six times. It also estimates that on average its 747-400s undergo around fifty wheel changes every year – which is almost one a week!

- Chock the wheels and fit the landing gear ground lock pins.
- Lift the axle with a jack until the tyre is clear of the ground. The axle can be jacked up at either end and, as the truck tilts, only the wheels at the end being jacked are off the ground.
- Set the parking brake and check that the PARKING BRAKE SET message is showing on the EICAS – electrical power will be required for this.
- Deflate the tyre, when all the air is out, remove the valve.
- Remove the hubcap.
- Undo the wheel.
- Install the protectors for the axle.
- Remove the wheel using a wheel-change dolly.
- Write on the tyre the reason for its removal.
- Before installing a new wheel take a good look at everything – the wheel, the tyre, the axle and the brakes – for any sign of damage. Remove any excess grease.
- Grease the wheel bearings and the axle. Remove any excess grease.
- Align the brake disks (the parking brake will have to be released for this and reset afterwards).
- Using the wheel-change dolly, install the wheel.
- Release the parking brake.
- Remove the protectors from the axle.
- Tighten the wheel.
- Inflate the tyre.
- Lower the jack until the weight of the aircraft is on the wheels.
- Install the hubcap.

1 The truck beam is jacked up. Because the truck tilts, only the wheel to be removed and its opposite number are raised off the ground. *(EI-AMC)*

2 Having removed the old wheel, it is moved out of the way. *(www.baa. com/photolibrary)*

3 Inspect the tyre, and all the other components, before fitting the wheel. *(www.baa. com/photolibrary)*

4 Load the new wheel into the wheel change dolly for fitment on the axle. The brake unit can be seen on the axle. *(www.baa.com/ photolibrary)*

5 Slide the new wheel onto the axle. *(www.baa. com/photolibrary)*

6 AND 7 When the wheel is in position, tighten it. *(www. baa.com/ photolibrary)*

8 Job done! *(Ian Black)*

landing gear, hydraulic system and engines may be replaced. The D check requires the skills of engineers, technical writers, quality control inspectors, avionics technicians, sheet-metal workers, and airframe and power-plant mechanics, most of whom are government-licensed. When cabin equipment mechanics, painters and cleaners are added, the number of personnel swells to well over 100 per day. Scores of others provide essential equipment, parts and logistics support.

Over time, in-flight vibrations, fuselage pressurisation cycles, and the jolts of thousands of take-offs and landings cause cracks in the metal structure of the aircraft. To address this problem aviation employs diagnostic principles similar to those used in the field of medicine. Both use such tools as radiology, ultrasonics and endoscopy to detect what the human eye cannot see.

For a conventional medical X-ray, the patient is placed between a sheet of film and an X-ray beam. To X-ray the landing gear, wings and engines, maintenance inspectors use similar methods. For example, a sheet of X-ray film is placed at a desired point on the engine exterior. Next, a long metal tube is placed inside the hollow shaft that runs the length of the engine. Finally, a pill of radioactive iridium 192 – a powerful isotope – no bigger than a pencil eraser, is cranked into the tube to expose the X-ray film. The developed film helps to reveal cracks and other flaws that may require that the engine be repaired or replaced.

During the D check, samples of the aircraft's fuel and its hydraulic fluids are sent for laboratory analysis. If micro organisms are found in the fuel sample, antibiotics are prescribed. To kill jet fuel bugs – fungi and bacteria that can get into fuel tanks through the air, water and fuel – the tanks are treated with a biocide, a form of antibiotic. This treatment is important because the by-products of microbial growth can corrode the protective coatings on the surface of the tanks. Fuel probes in the tanks can also be affected and thus cause the pilots to receive inaccurate fuel gauge readings.

As a result of normal wear, vibrations, and internal seal damage, fuel tanks can develop leaks. A supervisor asks his assembled D check crew, 'Does anyone want to be a "frogman"?' The joyless but necessary chore falls to John. Looking somewhat like a scuba diver without flippers, he dons special cotton coveralls, puts

on a respirator connected to a fresh air supply, and takes tools, sealant, and a safety light with him. Through a small opening in the bottom of the wing, he squeezes his way into the defueled wing tank, locates the source of the fuel tank leak, and seals it.

Built into the wings of the aircraft, the fuel tanks of a 747 are a maze of walled compartments connected by small openings. Fuel tanks are no place for the claustrophobic. A 747-400 can hold more than 57,000gal of fuel. This fuel capacity makes it possible to fly extremely long routes nonstop, such as from San Francisco, California, USA, to Sydney, Australia – a distance of 7,400 miles.

Three stories above the ground on the flight-deck, an avionics technician inspects a built-in test pattern display on the TV-like weather radar indicator screen. Pilots use this instrument to detect and avoid thunderstorms and turbulence that may be as far as 300 miles ahead of the aircraft. So when the pilot turns on the 'Fasten Seat Belt' sign, he may have seen turbulence on his radar screen. However, to prevent injuries, many airlines request that when seated, passengers keep their seat belts fastened at all times, even if the captain turns off the sign. Atmospheric changes in the form of clear air turbulence are often encountered before pilots have time to turn it on.

During the D check, safety equipment, such as life vests and emergency lighting, is checked or replaced. When a check of the passenger emergency oxygen system is underway, oxygen masks dangle like oranges on branches. Jet aircraft routinely cruise at altitudes of four to seven miles above the earth, where the oxygen content and the atmospheric pressure are insufficient to sustain life. How is this problem solved? The aircraft's pressurisation system draws in outside air and then compresses it. This air is finally supplied to the cabin at an acceptable temperature. If the air pressure in the cabin falls below safe levels, oxygen masks automatically drop from overhead compartments. The emergency oxygen is supplied to the passengers until the aircraft descends to an altitude where the emergency oxygen is no longer needed. On some aircraft, oxygen masks are stowed in passenger seatback compartments, not in overhead compartments. That is why it is important to pay attention to pre-flight passenger briefings, which identify the location of the oxygen masks.

A heavy maintenance check is also the time to install new cabin walls and ceiling panels as well as to replace carpets, curtains, and seat cushion covers. Galley equipment is disassembled, cleaned and sanitised.

Ready to fly

After 56 days of inspections, checks, repairs, and maintenance, the aircraft is ready to leave the hangar and resume flying passengers and cargo. Only a small fraction of the maintenance operations have been mentioned here. But before flying again, the aircraft may be test-flown by a special crew to ensure that all systems function properly. It is reassuring to consider briefly how much expertise and technology go into keeping the aircraft that you fly in a mechanically sound condition.

However, the best single tool in aircraft maintenance is said to be the human element – sharp eyes and alert minds. The trained personnel take their jobs very seriously. They know that poor maintenance can cause big problems. Their goal is to provide reliable aircraft that will speed you to your destination safely and comfortably.

(This description of a D check is reproduced with permission from an article written by a US Aviation Safety Inspector on the intricacies of a Boeing 747 inspection, which originally appeared in the 8 September 1999 edition of AWAKE! Magazine.)

BELOW Ready to fly again. Lufthansa -400, D-ABVW *Wolfsburg* (appropriately named and coded 'VW' after Volkswagen's headquarters at Wolfsburg) pictured on the gate at Cape Town ready for its overnight flight back to Frankfurt. *(Ian Black)*

Chapter Seven

Recycling the 747

Dismantling a retired Boeing 747 and 'parting it out' is a huge task because it is such a massive aircraft. Large or small, the recycling of scrapped airliners is big business. Estimates put the number of aircraft destined for the scrapyard over the next 20 years at 12,000 as airlines upgrade their fleets to more fuel-efficient aircraft.

OPPOSITE Jumbo junkyard. Boeing 747-400, F-GEXA, is at an advanced stage of being dismantled before the aluminium shell is finally cut up for disposal. Her interior has been completely stripped out and all four engines and APU have been removed, as well as the wing flaps and spoilers. *(All photos by Jonathan Falconer unless credited otherwise)*

155

RECYCLING THE 747

When a commercial aircraft is retired from revenue service its owner invariably sends it to an approved specialised aircraft decommissioning and recovery company, where the useful parts are removed and returned to the client for onward sale (parting out), and other materials are either recycled or disposed of. Pressure is growing to reduce significantly the amount of waste that goes to landfill.

The Aircraft Fleet Recycling Association (AFRA) is seeking to persuade owners of parked aircraft that through dismantling and recycling of the metal, plastic and composite materials of retiring aircraft, the aerospace industry can maximise the value of its ageing investments.

Air Salvage International (ASI) is one such aircraft decommissioning and recovery company specialising in parting out and disassembling end-of-life aircraft. Based in Gloucestershire at Cotswold Airport, formerly RAF Kemble, it delivers professional services to a client base that ranges from major airlines to multinational banks, parts companies and airports. ASI has extensive experience of all commercial aircraft types including corporate and light aircraft, and at the time of writing (2012) has dismantled almost 450 aircraft worldwide.

Founder and Managing Director Mark Gregory says: 'We have over 15 years' experience of dealing and operating with major

airports all over the world, and operate in strict accordance with the relevant authorities governing the disposal of aircraft and their associated hazardous materials. We can provide a client with complete accountability of their aircraft and parts by providing full administration and tracking from start to finish.'

The residual value of an aircraft or 'asset' is down to its age and the traceability of its parts. Traceability is the key that enables a client to recover the most money from parting out. Modern airliners can be worth up to 50% more when they are broken up for parts than they would be as complete, airworthy aircraft. Bradley Gregory, who is a partner in the ASI

ABOVE Other 747s await their fate at Kemble. This is MK Airlines' Boeing 747-200F, G-MKGA, in March 2012. When the cargo carrier ceased trading in 2010, the MK Airlines fleet included eight 747 aircraft.

LEFT Flanked by its smaller cousins, a Boeing 737 and an Airbus A320, F-GEXA sits in the dismantling compound at Kemble.

LEFT Viewed from overhead, the sheer size of the 747 can be fully appreciated when compared to the other aircraft in the dismantling compound.
(Keith Wilson)

business with his father Mark, says that 'every component of an aircraft is documented. It's because all the parts have an audit trail that makes them high-value assets. It's only when records have not been kept properly or have been lost that prices can be lower.'

Bradley gives an insight into the complex business of parting out and recycling aircraft.

'Air Salvage International is purely a service provider. We don't own any of the aircraft that we part out or scrap. It's the banks and private leasing companies who own about 90% of the world's commercial aircraft. Some of the

ABOVE Engines and the APU are among the first high-value assets to be removed from the aircraft. A Pratt & Whitney JT9D engine can achieve a resale value of about US$600,000 at 2012 prices.

RIGHT Before ...

RIGHT ... and after – F-GEXA shows the results of four months of careful dismantling.

FAR LEFT
If passenger seating
cannot be sold on it
will go to landfill.

LEFT Bradley Gregory
is partner with his
father, Mark, in Air
Salvage International.

larger airlines buy their own aircraft. A few of the smaller carriers do too, which are usually older models but with refitted interiors to make them look smarter.

'To give an idea of the costs involved a New Generation (NG) Boeing 737-600, 700 or 800 can cost anywhere between US$50 and $85 million new. We've been parting out ones at seven years old where the value of the parts is between US$30 and $35 million.

'Potential buyers make different assessments of a retired aircraft before they decide to make a bid. Out of every 100 aircraft that are looked at, only 2 will fit the buyer's requirements on price and suitability to part-out in order to make the money they want. In the end they will bid for only one aircraft.

'When dismantling we work from the aircraft's nose to the tail, then from the wingtips into the centre. All the high value parts are carefully removed by our qualified and European Aviation Safety Agency (EASA) accredited engineers – engines, APU, avionics, instruments, air-con units, the fuel system and its components. Among the last parts to come off are the landing gears, flying control surfaces, cockpit windows and external doors.

'From a Boeing 737 NG some of the highest-value parts are the thrust reversers which, with the right history, can fetch up to US$4 million for a set of four. On an older aircraft like a Boeing 737 Classic these parts don't fetch such great values, but the actuators fitted within can still be removed and retain a good value for their age.

'At 2012 prices a set of serviceable landing gears for a 737 Classic realises about US$75,000, depending on their age (number

ABOVE This is the skeletal port wing stripped of its outer flaps and engine.

BELOW The dismantled section of trailing edge flap with its actuators lying nearby.

of landing cycles) and traceability, while the same from a 737 NG is upwards of US$1m. The reason NG landing gears are so expensive is that there are so few in circulation at present, but it's still cheaper for operators to buy them from a parted-out 737 than new from the manufacturer. A pair of engines with the right paperwork can realise US$15 million.

'Of course dismantling a 747 and parting it out is a huge task because it's such a massive aircraft. Getting rid of its interior is usually the biggest headache for us. Seats and interiors on newer aircraft can usually be sold on, but with older 747s the interior fittings are at present not cost-effective to recycle and unfortunately have to go to landfill if the customer cannot find a buyer for them.

'Up until now carbon-fibre components have been problematic when it comes to their disposal, and on the newer model 747s like the 400 series they're used widely. It's a product used extensively in the aerospace industry to reinforce the class of materials known as carbon fibre or graphite reinforced polymers, and they've got a wide range of applications in an aircraft's structure.'

Great strides are being taken on recycling initiatives for carbon fibre by Boeing, AFRA and ASI who are working with the likes of University of Nottingham Faculty of Engineering to make the disposal of carbon-fibre composite materials more cost-effective. Using recycled or recovered carbon fibre instead of 'virgin' fibre (pure, primary or newly produced fibre) can also help reduce manufacturing CO_2 emissions by a reported 90–95%.

President of Boeing UK Ltd, Sir Roger Bone, said: 'The ultimate aim is to insert recycled materials back into the manufacturing process, for instance on the plane in non-structural sustainable interiors applications, or in the tooling we use for manufacture. This work helps us create environmental solutions throughout the lifecycle of Boeing products.'

Once the aircraft has been stripped of all the

RIGHT **Flight-deck instrumentation and fittings have been carefully removed to be reused.** *(Chris Wood)*

high-value parts, it's time for the disposal to begin. Bradley again:

'We hire a 35-ton 360-degree excavator with a grab and shears to cut up the shell. The first move is to bite into the back of the aircraft and allow the tail to fall off. We then bite forward all the way to the rear spar of the wing. At that point the operator changes attachments to shears and then cuts through one wing to pull it completely away from the fuselage. He then knocks the aircraft off the railway sleeper packs on which it has been positioned after removal of the landing gears. The next step is to cut up the wing. Root to first pylon, cut down the other side of the pylon and that will go; then to the next pylon and to the other end of the pylon.

'The toughest bit comes next and that's punching through the centre box. In fact it's also the dirtiest part of the job. The centre box comes away in splinters of metal or shards, unlike the fuselage that crumples up. Next move is to bite forward along the fuselage at mid-height, through the spar band that runs around the circumference to the upper deck. Then the shears work forward to the avionics bay and separate the nest of wiring.

'The aircraft has been cut into pieces leaving about 80 tons of scrap aluminium, which will fill between 10 and 12 trucks as bulk loads. The metal is taken by road to Robert Gibbs in North London, a scrap metal merchant accredited by AFRA, like ASI, where the metal is fragmented to get rid of any rubbish and then it's shredded before being melted down.'

The molecular structure of aluminium is not damaged by recycling, which means it can be reused almost indefinitely. So a scrapped Boeing 747 could easily become a boat, baking tin or a computer in its next incarnation.

RIGHT **An inglorious but environmentally friendly end awaits F-GEXA.**

Once you no longer have a need for your 747 there are several options other than scrapping. A large number of passenger and Combi aircraft have been converted to freighters. There are two main options: the -400BCF (Boeing Converted Freighter) and the -400BDSF (BeDek Special Freighter). Additionally Boeing designed another version for their own use, the -400LCF (Large Cargo Freighter).

747-400BCF (Boeing Converted Freighter)

The -400BCF is a modification of the passenger version, designed by Boeing but with the initial conversions carried out by TAECO at Xiamen in the People's Republic of China. Subsequently Boeing provided kits for other maintenance organisations to carry out the work.

Modifications included strengthening the wing to body joint, the main-deck floor and fuselage frames, adding a main-deck cargo door, adding a cargo handling system to the main-deck floor, and removing all the passenger equipment.

The first aircraft to be converted was a former South African Airways passenger aircraft, which was delivered to Cathay Pacific Airways in December 2005.

747-400BDSF (Bedek Special Freighter)

There are two options for the -400BDSF – it can be a conversion of a standard passenger aircraft or of the Combi. It was designed by Bedek Aviation Group of Israel, independently of Boeing, who also carry out the conversion work at their facility in Tel Aviv.

Perhaps not surprisingly the modifications are very similar to that for the -400BCF and include a main-deck cargo door, reinforcement of the main-deck floor and installation of a cargo handling system. In 2003 Bedek were claiming the cost of buying an aircraft and converting it to be in the region of US$54 million to $58 million.

The first aircraft to be converted was a former Air Canada -400 Combi, which was converted for Air China Cargo in 2006. The first passenger aircraft conversion followed a few months later with a former Singapore Airlines aircraft converted for Air Atlanta Icelandic.

The most obvious difference between these and aircraft built as freighters is that these aircraft retain their stretched upper deck. The only obvious external difference between the -400BCF and the -400BDSF is that on the

BELOW Cathay Pacific Cargo's 747-400BCF, B-HOU, takes off from Anchorage on 11 April 2008. *(Karl Drage)*

upper-deck the BCF still has some windows behind the upper-deck doors, whereas the BDSF has all its windows in front of the doors.

747-400LCF (Large Cargo Freighter)

The -400LCF is an outsize freighter designed specifically for Boeing to transport major 787 Dreamliner components to the final assembly plants at Everett, near Seattle, and Charleston, South Carolina. Consequently it has become known as the Dreamlifter. Although it was primarily designed by Boeing the conversion work was done by Evergreen Aviation Technologies Corporation in Taipei. Four former passenger aircraft have been converted; the first made its initial flight on 9 September 2006 and entered service in June 2007.

The aircraft underwent drastic surgery to turn them into Dreamlifters. The most obvious difference is the expanded fuselage which has a cargo capacity of 65,000cu ft, three times the volume of the standard -400F. The other major alteration is the addition of a 5ft-taller, hinged, vertical stabiliser. The tail swings open and the cargo is loaded through it, unlike the -400F where the cargo is loaded through the nose. As a result of some vibration issues discovered during flight-testing the winglets were removed. Internally it has a strengthened main-deck floor and a pressure bulkhead immediately behind the flight deck, which means that the cargo bay is unpressurised.

Because of the severity of the modifications, restrictions have been placed on their use and they are only allowed to be used by Boeing for its own operations.

ABOVE Asiana Cargo's 747-400BDSF, also at Anchorage, on 14 April 2008.
(Karl Drage)

BELOW LEFT Capable of carrying more cargo by volume than any other aircraft in the world, the Boeing Dreamlifter is derived from the 747-400 passenger aircraft. It is the primary means of transporting major assemblies of the Boeing 787 Dreamliner from suppliers around the world, to the 787 final assembly site at Everett, Washington State. This reduces delivery times to as little as one day from as many as 30 days today. This is 747-400LCF, N780BA, taking off from Anchorage, Alaska, on 5 July 2011.
(Paul Filmer)

Appendix 1

Boeing 747 technical specifications

A look at how the weights of the 747 have grown over the years shows not only how many different versions there have been, but also how much has been achieved with the original design. The original -100 series had a maximum take-off weight of 333.4 tons; with the -400ERF that had risen to 412.7 tons – an increase of 79.3 tons, which is almost 24%. (Another way of looking at it is that 79.3 tons is about the maximum take-off weight of the Boeing 737-800!) The -8I is up to 447.7 tons and is expected to go higher.

Dimensions

747-100, -200B, -200C, -200F, -300

Length	70.6m	231ft 10in
Wingspan	59.6m	195ft 8in
Height	19.3m	63ft 5in

747SP

Length	56.3m	184ft 9in
Wingspan	59.6m	195ft 8in
Height	20.0m	65ft 6in

747-400, -400F, -400M (Combi), -400ER, -400ERF

Length	70.6m	231ft 10in
Wingspan	64.4m	211ft 5in
Height	19.4m	63ft 8in

747-400D

Length	70.6m	231ft 10in
Wingspan	59.6m	195ft 8in
Height	19.4m	63ft 8in

747-400LCF

Length	71.7m	235ft 2in
Wingspan	64.4m	211ft 5in
Height	21.5m	70ft 8in

747-8F, -8I

Length	76.3m	250ft 2in
Wingspan	68.5m	224ft 7in
Height	19.4m	63ft 6in

Weights

MTW = Maximum Taxi Weight
MTOW = Maximum Take-Off Weight
MLW = Maximum Landing Weight
MZFW = Maximum Zero Fuel Weight

-100	kg	lb
MTW	334,750	738,000
MTOW	333,400	735,000
MLW	265,350	585,000
MZFW	238,816	526,500

-200B		
MTW	379,202	836,000
MTOW	377,842	833,000
MLW	285,762	630,000
MZFW	247,207	545,000

-200F		
MTW	379,202	836,000
MTOW	377,842	833,000
MLW	285,762	630,000
MZFW	272,154	600,000

-200C		
MTW	379,202	836,000
MTOW	377,842	833,000
MLW	285,762	630,000
MZFW	256,279	565,000

-SP		
MTW	318,876	703,000
MTWA	317,515	700,000
MLW	204,117	450,000
MZFW	185,973	410,000

-300		
MTW	379,202	836,000
MTOW	377,842	833,000
MLW	285,762	630,000
MZFW	247,207	545,000

-400		
MTW	397,800	877,000
MTOW	396,893	875,000
MLW	285,762	630,000
MZFW	246,754	544,000

-400D		
MTW	273,516	603,000
MTOW	272,156	600,000
MLW	260,362	574,000
MZFW	242,672	535,000

-400F		
MTW	397,800	877,000
MTOW	396,893	875,000
MLW	302,092	666,000
MZFW	288,030	635,000

-400 Combi		
MTW	397,800	877,000
MTOW	396,893	875,000
MLW	285,762	630,000
MZFW	256,280	565,000

-400BCF		
MTW	395,986	873,000
MTOW	394,625	870,000
MLW	295,742	652,000
MZFW	276,691	610,000

-400BDSF		
MTW	398,254	878,000
MTOW	396,893	875,000
MLW	295,742	652,000
MZFW	288,030	635,000

747-400LCF		
MTW	765,600	806,000
MTOW	364,235	803,000
MLW	*295,742*	*652,000*
MZFW	*765,600*	*806,000*

-400ER		
MTW	*414,130*	*913,000*
MTOW	412,770	910,000
MLW	295,742	652,000
MZFW	251744	555,000

-400ERF		
MTW	414,130	913,000
MTOW	412,770	910,000
MLW	302,092	666,000
MZFW	277,145	611,000

-8F		
MTW	443,613	978,000
MTOW	442,252	975,000
MLW	345,183	761,000
MZFW	328,854	725,000

-8I		
MTW	449,056	990,000
MTOW	447,696	987,000
MLW	312,072	688,000
MZFW	295,289	651,000

Appendix 2

Boeing 747-400 customers

In addition to their prototype, Boeing received orders for 724 of the Classic versions of the 747, which were made up as follows:

- 747-100 – 205
- 747-200 – 393 (including 4 E-4s for the USAF)
- 747-300 – 81
- 747SP – 45

Boeing built a total of 694 747-400s, as follows:
- Passenger – 442
- Combi – 61
- Domestic – 19
- Freighter – 126
- ER – 6
- ER Freighter – 40

This brought the total number of 747s built to 1,418, prior to construction of the -8 series. Customers for the -400 were:

- Air Bridge Carriers (3 ER Freighter)**
- Air Canada – 3 Combi
- Air China – 6 Pax, 8 Combi, 2 Freighter
- Air France – 7 (3)* Pax, 8 Combi, 2 (3)* ER Freighter
- Air India – 6 Pax
- Air Namibia – 1 Combi
- Air New Zealand – 4 Pax
- All Nippon Airways – 12 Pax, 11 Domestic
- Asiana – 2 (1)* Pax, 6 Combi, 5 (1)* Freighter
- Atlas – 15 Freighter
- Bahrain (1 Pax)***
- British Airways – 57 Pax
- Brunei (1 Pax)**
- Canadian – 4 Pax
- Cargolux – 16 Freighter
- Cathay Pacific – 17 (2)* Pax, 6 Freighter, 6 ER Freighter
- China Airlines – 17 Pax, 21 Freighter
- China Cargo – 2 Freighter
- China Southern – 2 Freighter
- El Al – 4 Pax
- Eva – 7 Pax, 8 Combi, 3 Freighter
- Garuda – 2 Pax
- GECAS – 1 Pax, 5 Freighter, 2 ER Freighter
- Guggenheim – 6 ER Freighter
- ILFC – 14 Pax, 1 Freighter, 3 ER Freighter
- Jade Cargo – 6 ER Freighter
- Japan Airlines – 34 Pax, 8 Domestic, 2 Freighter
- JASDF – 2 Pax
- Kalitta (1 ER Freighter)**
- KLM – 5 Pax, 17 Combi, 3 (1)* ER Freighter
- Korean Airlines – 27 Pax, 1 Combi, 10 Freighter, 8 ER Freighter
- Kuwait – 1 Combi
- Load Air – 2 ER Freighter
- Lufthansa – 25 Pax,*** 7 Combi
- Malaysian Airlines – 19 Pax, 2 Combi, 2 Freighter
- Mandarin – 1 Pax
- Nippon Cargo Airlines – 10 Freighter
- Northwest – 16 Pax
- Oman (1 Pax)***
- Phillipine Airlines – 3 Pax, 1 Combi
- Polar Air Cargo (5 Freighter)**
- QANTAS – 21 Pax, 6 ER
- Saudia – 5 Pax
- Singapore Airlines – 42 Pax, 17 Freighter
- South African Airways – 8 Pax
- Thai International – 18 Pax
- TNT (4 ER Freighter)**
- UAE – 1 Pax
- United Airlines – 44 Pax
- UPS – 8 Freighter
- USAF – 1 Freighter (YAL-1A)
- UTA – 1 Pax, 1 Combi
- Varig (3 Pax)**
- Virgin Atlantic Airways – 9 (4)* Pax
- Unannounced – 1 Pax

* Plus additional aircraft from leasing companies
** Ordered by leasing companies
*** Ordered by Lufthansa and supplied to other operators

Index